Social Media

Table of Contents

Introduction

When it comes to marketing your products, brand and services, you have to be as innovative as possible to remain on top of your game; given the amount of competition that exists out there, you have to be the best to make it big. If you settle for something mediocre or choose a beaten path, then you and your company will probably go unnoticed.

Ask any marketing expert about it and they will point to how it all boils down to making smart choices with your marketing strategies and trying to beat out the competition by being as innovative as possible.

It is of course easier said than done and you have to put in the effort to make all the right choices for your company. Given the plethora of options out there, it is obvious that you will need a little help; especially if you are your company's own PR manager.

One of the best, and most preferred, ways to advertise your products, and reach out to millions of customers worldwide, is by making use of Social Media. As you know, the Internet plays host to billions of people worldwide and you can easily reach out to many of them just by tapping into the different social media avenues.

In this book, we will look at how you can do so with ease and make your presence felt on all the different social media platforms. We will look at the individual media platforms in detail and understand why they are great choices for you and your company.

The main goal will be to beat your competition and stay ahead of the herd.

I thank you for choosing this book and hope you enjoy reading it.

Chapter 1:

What is Social Media? Why is it Useful?

For all those that are new to the concept of *social media,* we will look at its meaning and uses in detail.

What is social media?

Social media refers to various platforms that are available on the Internet, which provide users the chance to create their profiles and share and promote content. These social media platforms are all designed to help people and companies establish a social presence and let others know about their products and services.

The popularity of social media, as a marketing tool, rose in the last decade as more and more companies realized its true potential and began using it to their advantage. They understood that it is possible for them to reach out to millions of people, worldwide, and increase their customer base by several folds.

Social media is now part of every company's marketing strategy. Right from a small store in Japan to a multi-million dollar company in the US, everybody is using the power of the Internet to get noticed and improve their product's sales.

Through the course of this book, you will understand the real use of social media in terms of your marketing strategy and why it is extremely important for you to have a strong marketing plan in place in order to promote your products and services.

Let us now look at why social media is useful.

Understanding how social media marketing works

Many people, particularly small businesses and upcoming businesses seem to shy away from social media marketing. The reason being that it social media marketing is such a vast area and it is really difficult to know where they can start from, the area that should be targeted by them and their potential audience. In order to become successful b making use of social media marketing you really need to get yourself better acquainted with the complex hierarchy that exists and the workings of these platforms. Like it is the case with any online marketing strategy, even in the case of social media marketing it will all have to start with you and your website. Your website will act as the foundation, the base on which you can start building your campaign. Blogs have been gaining a lot of publicity, if your website has one then that's good and if it doesn't then perhaps it's time that you have added one. Blogs will enable you to provide regular steams of content that will help you grab and then hold on to the attention of your audience and in order to increase the number of people who are subscribing to your website, increase the number of followers you have got on any social media platforms and for also giving your business a sound online presence, a blog will really come in handy.

You probably would have come across the term RSS feeds and you might have even used them. If you haven't done so, then you probably should. RSS stands for Really Short Simple Syndication. RSS is really a great tool that can help you sort your content out in a way that provides you with the option of personalization. The benefit of personalization is that you can sort your content according to different sectors and areas of interest of your audience. People can always subscribe to your website and they can always agree to receive the RSS feed that will let them view the content as well as let them know of any updates. When you can personalize the information available you can filter it depending upon the requirements of your target audience.

Then there is the option of bookmarking as well as social sharing. This can be thought of as the process that lets you tag people and also keys you share certain elements of the content you have got on various social networking sites such as Facebook, Twitter and even Google+ or even on social bookmark sites such as Delicious, Digg or even StumbleUpon. If you really want this to work, then you will have to ensure that the content that you are wanting to post is of high quality and is relevant to the audience. It really won't work if you are just sharing silly memes or GIFs. If your audience doesn't think the content that you are posting is relevant then it is highly likely that they will stop following you.

Social search tools will also be really helpful. Google Places, Foursquare, Yelp and even Bing Places are some of the popular directories that are available online and you can get yourself listed on these sites. So that when anyone is searching for your name then your listed address would come up on the search. This will help you draw more attention to yourself and will help you in strengthening your online presence which is really crucial for your business. Social search tools are highly recommended by me and this will help in acting as a catalyst for improving the publicity for your business.

The top social media platforms to make use of

There are very few people who wouldn't have heard of Facebook or Twitter, but these aren't your only options there are a variety of social media platforms that you can make use of. If you can make use of it in the right way then your marketing campaign will be successful.

It is a unanimous opinion that **Facebook** is the most popular social media platform. There are more than 1.2 billion active users on Facebook. This platform provides you with an opportunity to advertise about different kinds of businesses it can be paid or free advertising. It also provides you with the option of creating pages dedicated solely to your business and this can help you engage your potential customers. Facebook ads work according to PPC model that allows you to target certain ads and specific audiences. You can also share your content and communicate with your audience on a personal level.

Twitter is all about Microblogging and it is a networking platform where there are more than 200 million active users. This is considered to be a very popular platform for businesses, celebrities and entrepreneurs alike. Twitter users can post updates and these are known as tweets. A tweet cannot be more than 140 characters long and this condition gives twitter the feel of an SMS system. You can create your business stage and you can make use of this for attracting your customers as well as getting updates to your audience without much trouble. You can also make use of the promoted tweets feature that gives you access to paid advertising and you can reach a much wider audience by making use of this.

Google+ has more than 540 million active accounts and this is considered to be the second largest social media site in this world. It is fully integrated with a lot of other services that Google offers and it is a really good option for businesses as well as individuals who are looking for a platform for their soul a media marketing strategy. This is a professional platform and it aims at businesses by allowing them to form relationships with their customers, investors and other interested parties. Your profile on Google+ will be linked to all other Google services that you make use of such as Google places.

LinkedIn has more than 270 million active users and this is aimed for businesses alone. This is an incredible platform for anyone who is associated with the business world. Unlike all the other social media networking sites that cater to both businesses as well as individual users, LinkedIn solely caters to different businesses. Using this platform you can create a Company Page and this gives you the opportunity to showcase how well your company is doing and it also gives you a means of reaching out to your potential customers. This really should be your go to website if you want to develop your business connections, especially if you are involved in Business-to-Business marketing. This will allow you to find as well as hire employees or even search for any business leads by going through different profiles of likeminded people or people with similar interests.

Pinterest is not just a website where people get to pin photos. It is so much more than that. This is a unique platform it is not like any other social media sites and also social media marketing. This has around 70 million active users at present, making it a relative teenager in terms of its popularity but that number has been increasing at a steady pace. This is the place for both business users as well as individuals particularly those who make use of a lot of visual media. This would be for businesses that are related to the fashion industry like jewelry designers, photographers, any designers or basically any business that heavily relies on visual media. Pinterest also offers business accounts that come with added features that will let you analyze your pins and also help in promoting particular pins. Your business profile can also be easily synced with your accounts on other sites such as Facebook or even Twitter.

Instagram is another popular social media site that is aimed at social media; it can be either in the form of photographs or even videos. This site has over 150 million active users and it has become the latest fad irrespective of their age. This is a perfect option for all those businesses that rely heavily on visual media like fashion businesses, food, design, and travel and so on. Businesses can opt for either posting photos or videos of their products on Instagram and they can organize different photo or video contests for spreading publicity about their business. You can link your Instagram account to a business website but you can mention the same in any of your posts on other social media sites as well. You can make use of Instagram for generating web traffic for your website and thereby generate more interest.

In a strict sense YouTube really isn't a social media site but then it is the most visited video sharing site in the world and also the third most frequently visited site as well. **YouTube** combines a lot of features that make this site a vital tool to make use of in your social media marketing strategy. This is a free platform that anyone can access and you can make use of this site for publishing any videos that are related to your business or area of interest. You can also make use of the feature of paid advertising for promoting your products or services and this means that your ads would show up on videos that are posted by other users as well.

Why is social media useful for marketing?

As you know, social media holds a lot of potential and will help in increasing your customer base. You will see how easy it is for you to reach out to different customers that are based all over the world. Let us now look at some of its real uses to companies.

Reach

The first and most important use is the reach that this platform provides to its users. You will see that it is possible for you to reach out to more people just by making your presence felt on these sites. When you add one person, you will automatically end up adding another 10. This is not possible when you advertise in the traditional way. You will hardly be able to reach a few hundred there whereas here, you can easily reach millions just by clicking a few buttons and uploading pictures of your products and services.

Recognition

Your brand recognition will grow in leaps and bounds. Imagine having a small shop in a remote island and trying to reach out to the world. It will seem like a herculean task. But now, you can easily reach out and have our brand successfully recognized by million just by setting up an account on a social media platform. It is like getting to set up free billboards on every street in the world. Your brand is sure to be recognized by millions around the world and you will see that it is possible for you to become a global image by establishing your presence on social media.

Costs

The costs of marketing can be considerably reduced when you take up online marketing. When you market in the traditional way, you end up spending a lot of money. Right from paying the advertising company to paying for the different promotional campaigns, there are many costs that will keep accumulating. You have to set up a big fund for it and only then will you be able to afford the traditional method of marketing. However, with social media, all of that can be reduced to a bare minimum. You will see that it is possible for you to promote your products and services with a very small to no budget at all! Imagine the kind of money you can save on just by adopting social media for your marketing needs.

Interaction

Through social media platforms, you can interact with many people including your customers and potential business partners. You can bring everyone under the same roof and allow them to interact with each other. You will see that it is easy for you to answer any queries that these people have towards your company, products or services and make it an interactive session. This type of a setting will go a long way in helping you establish a good connection with your customer base.

Conversion

Through social media, you can easily convert people into your customers. Now say for example 500 visit your page on a daily basis. Out of those, maybe 200 are your existing customers and the rest are new people. If even half of them, meaning 150 people convert into customers then you will now have 200+150 customers for your products. That is a great number for you to work with, especially when you are just starting out. That number will only grow over time and before you know it, you will have a big audience base following you.

Loyalty

You can establish customer loyalty by being online. You can be in touch with the best customers and get them to be your repeat customers. Social media helps in establishing a strong customer base and also strengthen your hold over them. Remember that your current customers are extremely valuable as they are who will bring new customers and also give you consistent business. So it is extremely important for you to hold on to your loyal customers and get them to help increase your business.

SEO

You can make use of SEO when you set up an online account. SEO stands for search engine optimization, which will help you get noticed. You can use it to your advantage and turn up as the first search result online. That will ensure that your page gets visited more often, which will mean more customers for your business.

These form just some of the benefits of using social media for marketing but are not limited to just these. As and when you start using social media you will be acquainted with the other benefits. Following are the likely benefits and the reasons why you should start making use of social media marketing right now.

Helps in improving your brand authority:

You shouldn't forget the basic of marketing just because you are making use of social media. You will have to keep interacting with your customers on a regular basis and when you do this it shows good faith towards not just your existing customers but your potential customers as well. With the advent of technology people these days like to brag or even criticize about a particular service or product on various social media platforms. And when they do happen to post about a particular brand online, they are in fact introducing this brand to several others on an indirect manner and this expands your audience base. As the number of people who are talking about your brand starts to increase then the publicity of your brand will increase too and it will be perceived as being more valuable. You can always tie up with individuals who are quite popular on social networking sites for the promotion of your product. This will help in increasing the publicity of your product manifold.

The inbound traffic increases:

Your inbound traffic is generally restricted only to your existing customers and all those users who would have searched for the keywords that currently work for your product or brand. But social media can really help you turn things around. Every profile that you manage to add on social media will help you create a pub that will directly lead to your website and you every piece of content that you have managed to acquire I'd an opportunity for attracting new customers. When the quality of the content that you make use of or publish on social media strata improving then it will also help in increasing the chance of generating conversions will also improve.

Reduction in the marketing costs:

According to an online report published by Hubpost, approximately 84% of all the marketers had to put in just around six hours every week and they had managed to generate a noticeable increase in their web traffic. Six hours is comparatively a very small price to pay for the more than proportional increase in your brand recognition. A little bit of effort can help you reap the benefits of social media marketing. Even if you are able to spend maybe an hour everyday for developing your content as well as designing the strategy for marketing, you will be able to see results in no time. The option of paid advertising can always be opted for, but whether or not you want to make use of it solely depends upon your goals. You needn't worry about started out small, it needn't be on a large scale and you needn't exceed your predefined budget. Once you have acquired an understanding of how social media marketing works, you can slowly start increasing your budget according to your needs and you will definitely be able to improve your conversion rate.

Better search engine rankings as well:

SEO can be thought if as one of the best and simplest manner in which you will be able to capture traffic that is relevant to your content and then direct such traffic towards your website. The requirements of this mode keep on changing constantly. It is not just about regularly updated the content on your blog, the optimization of the titles used and the distribution of such links that all lead back to your website. Most of the search engines tend to make use of social media presence for calculating their rankings and most of the established brands also tend to make use of social media in one form of the other. Just being active on social media is sufficient to send a signal of credibility to the search engines regarding your brand. To put it in a nutshell, if you really want your brand rankings to go up then you will need to have a really strong presence on various social networking platforms.

Better customer experience:

Social marketing is a channel of communication that really isn't much different from the traditional channels of communication such as phone calls or even emails. Every interaction that you might have with a customer on social media should be considered as an opportunity for you to promote your brand and you can do this by projecting a good customer service experience and it also facilitates in helping you to enrich the existing customer relationship. If at all a situation arises wherein a customer has taken to Twitter to express their grievance about any particular product then you should be able to take an immediate action and rectify the problem as well as apologize to them in the same public forum. But not all of your experiences are going to be negative. If any customer expresses their satisfaction and happiness with your product then you can express your gratitude to them and you can also provide them with a list of additional products that you would recommend. You should make use of social media for improving personal communication with your customers thereby providing them a personalized experience.

Improvement in customer insights:

Social media can prove to be really helpful because it also provides you with an opportunity to understand how the customers behave and this can be done through something that is referred to as social listening. You can do this by opting to monitor the comments that your customers might post; this would give you an insight into their personality and what they think of your business. Another thing you can do is opting to segment the syndicated content and this will let you understand the content that has been able to generate the most interest and according to this you will be able to post further related content. Not just this but you can also measure your rate of conversion depending upon the different promotions that you managed to post on different social networking platforms. Being able to determine the most used social media by your customers will let you understand the media channel that you should actually make use of. You can really make a move to improve your revenue if you know what exactly your customer wants.

The above-mentioned are the benefits that you can derive by making use of social media marketing. Bet if you really aren't fully convinced even now about making use of it; then here are some other things that you should really consider before writing it off.

Your competitors are already involved:

You need to understand that your competitors might already be engaged on various social networking platforms and they are making the most of social media. So, you really wouldn't want to miss out on any social media traffic. You should also realize that since your competitors are already involved in social media, they are poaching your potential conversions. You should also get going and not remain like an idle bystander anymore while your competitors are enjoying all the benefits of social media marketing. Leveling your playing field must be reason enough for you to give social media marketing a whirl.

The sooner the better:

The foundation on which social media marketing is based is relationship building and you can always do this by trying to expand your followers. This will help you attract more and more customers. The sooner you start the greater will be the number of audience whom you can attract.

Needn't worry about potential losses:

If you really think about it, then you will realize that there really won't be any losses that you will be incurring. The amount of time and money that you will be spending will be an insignificant fragment when compared to the potential profits that you can make. You don't necessarily have much to lose by making use of this but you do have a lot to gain. All you need to do is put in a few hours of work and spend a couple of hundred of dollars for getting started. This is all the investment that you need to make and you will definitely be able to reap way more than you will have to invest.

You really shouldn't be waiting any longer and should get started as soon as you can. The more you wait the higher are your chances of losing out on potential business. Social media marketing can really come in handy and help you attract a lot of customers and it can also help in improving your conversion rate. So, all you need to do is get started. Take the first step and jump onto the bandwagon, because whatever the cynics might say, social media marketing is here to stay.

Chapter 2:

Getting Started with Social Media

I t is important that you get started with social media by doing all the right things. In this chapter; we will look at the different steps that you must adopt to start with social media on the right foot.

Research

The very first thing to do is conduct a timely research on the topic. You have to seek information on the topic of social media marketing and remain as educated and informed about it as possible. This book will act as your one true guide and give you enough information on the topic. You will be able to start with your online account by the time you are done reading this book. But you must also turn to online sites for information and other books from good publications. As long as the information is genuine, it will help you in a big way, and assist in getting started.

Features

The next step is to understand the different features of each of the social media platforms. Each one will have a different feature to offer which you must understand in order to pick it. You will see that it will be an easy process for you to pick the right platform once you understand what each of them has on offer for you. We will look at it in detail in future chapters of this book and it will help you make your choice.

Choosing platform

The next step is to choose the best platform that you can use to set up your online presence. It is a good idea to pick all, as that will help you connect with a larger audience. There are four main types of social media platforms namely Facebook, twitter, YouTube and Instagram and all of these will help you in their individual way. But if you wish to dominate in just any one then you must pick the best one. That, you will be able to do, only after you go through all the features of these platforms!

If you decide to concentrate your efforts on the most appropriate platforms for your business, how do you go about it? Did your parents ever remind you to choose your friends wisely back when you were a kid. When it comes to choosing the best social media platforms for marketing your business, keep the same advice in mind. Because there are many social media platforms that are available out there, it's important that you be able to pick only the best platforms to market your business so you can maximize your limited resources and time.

Don't just follow the social media marketing crowd blindly. When it comes to social media marketing, you're better off taking the same approach in choosing your hangout spots when you were still a kid: asking yourself "Where do the people that I like to be with, i.e., the cool kids, hang out?" And as you become with social media marketing for your business, you'll see that the terms "relevant" or "cool" are highly subjective and means a lot of different things for certain people.

For effective social media marketing, you must learn as much as possible about your customers and prospects, especially what they are most interested in. As you do, you will need to narrow down the social media platforms that your customers and prospects are most likely to use.

Here are some of the most popular platforms for social media marketing today as well as what types of businesses can most benefit from them. You can use this as a guiding principle in choosing your business's particular social media platform.

Facebook

If you're just starting to build the social media presence of your business or if your main goal History Channel people as possible and social media, Facebook is obviously the right platform for you. Although it's reported to be losing grip on many of its younger users, more than 70% of adults you are always online are still participating actively on Facebook, which makes it still the biggest and most popular social media platform today.

Aside from just having the most number of active members or users, Facebook is also the social media platform that has the highest engagement levels as indicated inept status as the most frequently used platform.

The only potential drawback or limitation when using Facebook as platform for social media marketing is it primarily relation of nature, i.e., it's used mainly for personal relationships with friends family and colleagues. In this your primary objective for using this platform is to engage people with your business, there's a small possibility that it may not necessarily be the best or most effective platform to promote your business and social media.

Instagram and YouTube

If your products, services and customers are the type that are highly visual, Instagram and YouTube maybe the best social media platforms for your social media marketing campaigns. Particularly with Instagram, it interesting to note that it can overlap with Twitter and Facebook and as such, you can use Instagram as part of an excellent triple-edged social media sword.

Instagram is a great platform for social media marketing especially for targeting key niches because its appeal extends to specific ethnic segments and is very popular with people located in highly urbanized areas.

Now, Make That Choice(S)

After reading about the most popular platforms for social media marketing, it's time to decide. Such a decision or decisions include looking for prospects, being where they are most likely and posting the right content, whether it's someone else's (with proper acknowledgment of course), yours or simply shared from other trusted sources like news, sports and entertainment websites, among others. Such decision making also means you must be realistic in terms of what you plan to do as well as how much to do when it comes to interacting or engaging prospects and customers. Keep in mind that social media is one that's reciprocal that demands your participation in what prospects and customers deem to be as sensible and relevant conversations.

Setting up

The next step of the process is to set up your social media platform. You have to work on your profile, your display picture, upload pictures of your products, write descriptions etc. All of these will help you set up the best profile for your products and services and help your audience connect with your brand. You have to take your time when you indulge in this step. Don't be in a hurry to set it up and be done with it. Regard this as one of the most important steps of the process and pay keen attention to all the small details of the step.

Target audience

The next step is to understand who your target audience is. You have to cater to them if you wish to increase your sales. You have to make the effort to find out who your target audience is. It can be just a few or many depending on how many products and services you have on offer at your company. Your target audience should be grouped in certain groups so that it is easy for you to cater to them individually.

Promotion ideas

The next step is to have your promotional ideas in place. These ideas will help you reach out to a lot of people and get them to like your products and services. You have to think up unique things that are not what most of the other companies are adopting. Try to keep it simple yet intriguing. Everything should be appealing and capture your customers' imagination. With time, your promotions will start getting better and your audience will start buying more and more of your products. You have to learn to back link all your products with your website so that people can visit the website and buy the products.

Market research

The next step is market research. You have to conduct timely research on your products and services and how your audience perceives them. You have to see if they like your promotional activities and if it is getting across the desired message. You should ask them questions and get them to answer them to the best of their knowledge about your products and services. Once they give you the feedback you can make the changes in your marketing strategies and satisfy your customers' needs.

Timing

It is always important for you to time your entry into the market and also your promotions. You should see when it is a good time for you to promote a certain type of product. You have to look at when your competition releases a product and time yours accordingly. In fact, you should promote it well in advance and release it slightly before they release theirs.

Goals

The next step is for you to set goals for yourself. Ensure that they are reasonable and gettable goals that you can easily attain. Try to set a new goal each time a previous one is attained. You have to write down your goals and tick them off one by one.

These form the different steps that you must follow when you wish to set up a social media presence. These have been mentioned in a stepwise manner and you must follow the same if you wish to see quick results. But if you have a plan of your own in mind then you can follow it by using this as a guideline.

Chapter 3:

Master Facebook

Facebook is the world's biggest social networking website with more than 1 billion users that login on a daily basis! Not just that, it also invites a billion likes a day and millions of comments on the different posts. Consider too, the following statistics as of 20 September 2015 from statisticsbrain.com:

- 1,440,000,000: total number of active users monthly;
- 874,000,000: Total users of Facebook on mobile;
- 12 %: Increase in members in 2015;
- 640,000,000: Total minutes spent monthly on the platform;
- 48 %: Percentage of total users who log on to their accounts daily;
- 18 minutes: Average duration of Facebook visits; and
- 74,200,000: Total Facebook pages.

So it would be a no-brainer for you to have a Facebook account in lieu of your company if you wish to reach out to billions.

Facebook is an easy to use website that is also fun to navigate. All it takes is a few minutes for you to register and you will be lead to your home page. There, you can customize your profile and make it as attractive as possible. It is important that you personify your company and make it look like your company is operating the page.

Once you get started, you must send the link of the page to all your contacts so that you can add them in your friend's list. You have to inform that the page belongs to the company and ask them to contribute to its growth.

Rest assured, you will be able to amass quite an audience within no time at all. After all, the odds will be in your favor when you are working with numbers that run into billions. Within no time will you start to see that there are hundreds that are adding you and liking whatever you are posting on your homepage.

It is extremely easy for you to reach out to your target audience through Facebook. You can club them together into the same groups and send out-group messages to them.

You have to keep it as interesting for your audience as possible. Try to use all the best quality pictures and give a proper description of whatever you are uploading. All of it will go a long way in helping you amass a large audience. You have to be as transparent to them as possible and keep them informed about everything that is ongoing in your company.

Don't keep it to just pictures and text; you also have to add in videos and gifs to make it interesting for your audience. They have to feel attracted to your page in order to visit and use it.

Once you set up a Facebook profile for your company, you have to move to the next step, which is to create a page for it. Let us look at that aspect in detail.

Why Facebook Marketing?

While it's certainly not the case that all 1 billion ++ plus Facebook users can be your actual prospects and possible customers, its huge membership base implies that there's no other social media platform on the face of the earth that's better than Facebook in terms of exposure. But wait, there's more! This is just the tip of the iceberg and let's take a look at why Facebook marketing just makes so much sense.

Segmentation

Here's a cool piece of trivia info: did you know that Facebook contains a huge, huge database on practically anything and everything that can be related to all of its users? Interests, location, favorite things, age and more ... talk about intimacy! Now, what am I driving at here?

There are two ways that you can advertise or promote your business on Facebook: for free and through paid advertisements. The database or the information that I mentioned a while ago is of particular use for paid advertising. Why?

The usual or traditional way of advertising such as TV, print ads or radio employ what is called a shotgun approach, i.e., mass advertisements with very little opportunities to choose the audience that can view the ads. Unlike traditional advertisements, Facebook gives you the ability to focus your paid advertisements on specific niches or markets.

An example of this would be a Japanese restaurant located in Nevada. Let's assume that this Japanese restaurant is rather unique in that it wants to focus specifically on people who love Japanese food who practically worships K-Pop artists. This restaurant can be advertised on Facebook and target it to Facebook users who live in Nevada, who love K-Pop music and who gorge on Japanese food.

Nevada however, is a very large area. As such the owner of this restaurant would be better off narrowing down further his or her target audience to those living in say, Reno. That's how specific or targeted Facebook advertisements can be.

Let's say you're a public speaker who specializes in giving talks and seminars on relationships, particularly marriages. You can market your services or your upcoming talks and seminars on Facebook with paid advertisements and target people who are between 25 - 35 years old who would like to learn more about how to make their marriages beautiful or better. And if you are a bit chauvinistic, you can even limit or target the advertisements to the husband.

Facebook advertising can be that specific – segmented – compared to traditional advertising.

Cost Management

It's relatively easy and practical to keep your advertising budget under control with Facebook marketing campaigns. You're not just able to control the maximum money that you'd like to spend for campaigns, but you can also control how long those campaigns will run for and how much to spend on them every day.

Say for example you only have a budget of $30 tops for a Facebook marketing ad campaign that you want to run for 30 days. Given those parameters, Facebook automatically puts a maximum spending amount of only $1 per day on your advertising for the next 30 days to help you stay within budget. With Facebook advertising, you need not worry about mindlessly blowing your budget.

Faccbook Pages

Facebook gives you the option of creating a separate *Page* for your company. This means that you can create a page and send the links out to people to like and follow. This is unlike a regular profile that you create, as there is no adding of people. You can create a page to use as a platform to tell your customers about your products, schemes and promotions.

Here is a step-by-step guide for you to get started with it.

Step 1

The very first step of the process is to visit the homepage. There, you will be able to create the page for your company.

Step 2

It is important that you fill all of the fields there and not leave anything out. That is what will make the difference between a good page and a mediocre one. If you leave something for later then it will never get done. Add in all the important keywords in and ensure that people looking for it easily find your page.

Step 3

Choose the correct category for your business in order to make it easy for the people to look for it. You should look at the categories that are mentioned and choose the right one. That will make it quite easy for anyone to find you.

Step 4

The next step is for you to customize your URL such that it tells the reader whose page it is. You company name should be present in the URL so that people immediately know where they are being redirected to. You have to put in efforts to make it as easy for your readers as possible if you wish to make the most of your Facebook page.

Step 5

Start posting all that you wish to. If you are a bakery then you should upload the menu, post pictures of your products, inform them about any customization that you will do, tell them about your special menu etc.

Step 6

Now, don't be in a hurry to invite people to like you. Ideally, you must wait a day to fill everything out and see if you have to make any changes. You must also ask a few friends and family members to look at the page and suggest any changes that you might have to make in it. Once everything is sorted out and you have made the changes, you can start sending out invites to people.

Step 7

The next step is to add a *Like us on Facebook* button on your website. This is important, as you have to tell your audience that you exist on Facebook. Once done, you have copy and paste your URL in all your other social networking sites if you are already active. You should also place a link as your email signature so that people will know to click on it.

Step 8

After some time has passed, you have to introduce interesting content to keep people's enthusiasm going. If you started with pictures then you should start adding in videos to educate the people and also some demonstration videos for them to see and enjoy.

Step 9

Employ a Facebook specific team that will look into keeping your Facebook updated. This team should comprise of people that know exactly what they are doing. Try to employ those that have relevant experience in this field. After all, they have to promote your business and help it reach heights. So there is no point in compromising on this step and you have to employ the best. You have to instruct them to post new content at the exact same time. Even if it is posted on alternate days, they should get the timing right. That will create a sense of responsibility in them and your audience will know exactly when to check your Facebook page out. You should also instruct them to limit the size of the posts to 250 to 350 words only as research has shown how that is the best length for Facebook posts. Get them to patiently reply to all your audience's queries and remain as interactive as possible. The whole point is to be interactive and tell them things that they don't already know about your company and products. Satisfy their need for information and your company will do really well.

Step 10

You have to strive to increase your traffic. This you can do by posting interesting content that your visitors really want to read. When they will see what they like on the search engine, they will definitely visit your page. For that, you have to think from your customers' point of view. That will tell you what to use as an advertisement to attract your viewers.

Step 11

You have to join as many diverse groups as possible in order to promote your own page. When you like another group then you can educate them about your page and get many people from there to like you back.

Step 12

There is the option of having paid ads on your page if you wish to garner traffic. This is only if your page is not doing as well as you want it to. The paid ads are a great way to generate traffic on your page. It is a powerful strategy that many companies use to have their pages liked by many people o simply generate unique hits.

These form the different steps that you must adopt in order to set up your Facebook page. Remember that updating the page from time to time is vital and you have to give your audience something new to look at every single time.

The Art of Facebook Marketing – The Porch Analogy

You'll need to learn about Facebook's marketing practices if you want to optimize the use of its features as well as your market reach. Although these practices are not considered to be unbreakable rules, following them will help you substantially increase your chances of engaging your customers well on Facebook.

Social media marketing is primarily about engaging people. If you check out many of the world's most popular business Facebook pages, one very obvious aspect to them is that they hardly ever do direct selling to their followers and they don't preach to their markets. What they do instead, is engage.

Another thing going for the market today is that most people are a whole lot more cognizant and knowledgeable about marketing and products to the point that they can quickly recognize if you're just trying to get a quick buck from them or if you're genuinely and sincerely relating to them. Always be cognizant that Facebook is first and foremost a social network and not the sales and marketing platform.

In one of the best books on Facebook marketing called The Ultimate Guide to Facebook Advertising, expert authors Perry Marshall and Thomas Meloche explained clearly, using a front porch analogy, just how different Facebook marketing is from the traditional advertising that we're so familiar with. They explained it this way, though I took the liberty of rephrasing their analogy:

> *Think for a moment that you live in the town square.*
> *Further, see in your mind's eye that the house you*
> *live in has a front porch where you hang out often*
> *and enjoy looking at people passing by as well as*
> *cultivating many colorful and attractive plants.*
> *There are times when, as you look at people and*
> *drink from a cold pitcher of lemonade, that some*
> *passersby see the beautiful and colorful plants on*

*your front porch and come to you to inquire about
how you're able to keep them looking so.*

*As they do, you offer them to sit down on your porch
and poor glasses of that deliciously cold lemonade
for them to drink as you explain in general terms
how you keep your plants looking healthy and
beautiful. With your explanations, some of the people
who you talked to have become so intrigued about it
to the point that they want to spend the day with you
just to learn your secrets - and they don't mind
paying for the opportunity to do so.*

*Thinking that you enjoy telling people how to keep
their plants looking healthy and beautiful as well as
this is a good opportunity to make extra money, you
take them up on their offer. And so the next day, you
find yourself spending the whole day teaching them
the fine art of cultivating beautiful and healthy
plants.*

Now tell me, did you notice any direct or explicit attempts to sell stuff on the porch? Or how about any implicit or subtle attempts to promote an ulterior motive, which is a horticulture seminar? I bet your answer is no. And that's right. That's a very good bird's eye view of how effective Facebook marketing is done. Selling, if any, is done within the context of personal connections and relationships only.

I never get tired saying this but social media marketing is primarily about engaging other people. And one of the best ways to do that on Facebook is by posting useful tips and links to other materials that they like and will probably share, as well as by asking them questions that are relevant and sensible. Focusing your posts on your audience instead of your brand help you develop relationships by addressing their needs and piquing their interests. Remember, relationships are the single biggest reason for why social media was created and as you saw in the porch story above, sales is just a fruit of such relationships.

Another excellent method for engaging your customers and prospects on this giant social media platform is by regularly posting content every day. And I'm not talking about regular, ordinary content. I'm talking about unique and the high quality ones. While it seems like too much work to post such content frequently considering you're very busy schedule, not doing so significantly increases the risk that your target prospects or customers will miss one or two of your important posts. And if this continues, they'll tend to lose interest in your business' Facebook page and "like" or "follow" the Facebook pages of other businesses. And as they like more and more Facebook pages, the competition for their attention and engagement becomes even more intense. As such, you can expect them to completely drift away from your grasp. And there goes your social media campaign.

When we talk about posting regularly, how regular is regular? Well, there are studies that recommend between 3 to 5 posts everyday for optimal engagement with customers and prospects on Facebook.

Each and every situation can be different however and as such, I recommend that you strategically and sensibly experiment on the frequency that best works for your business's social media campaign.

Finally, you should make your posts interesting and fun. Remember that a huge part of being able to successfully engage your customers and prospects involves everyone having fun. As such, keep your posts helpful, useful and relevant but interesting and light at the same time, whenever possible.

Knowing Your Target Market

Some people are very concerned about the vast number of demographic information that Facebook keeps about its users. But for social media marketers, it is something to bask in.

But just how specific can the demographic information the Facebook keeps on its users be? Consider that among the many types of information Facebook has, it includes your location, age, gender, interests, work, political views and life events. And as I mentioned, if you're into social media marketing, this is something that's very desirable.

Keep in mind that when it comes to social media marketing, the more you know about your prospects and customers, the more effective your Facebook marketing campaigns can be. This is because you'll be able to structure or design your advertisements in such a way that will grab their attention, because their interests, engage them well and even convince them to patronize your business.

But while general demographic information on Facebook gives you general insights about your audience, you can never truly know each of them individually – at least not completely and perfectly. In other words, there will always be some blind spots or gray areas in terms of knowing them. But of course, that shouldn't stop you from conducting social media campaigns on Facebook using engaging and interesting posts.

Always remember to make sure that each and every post on your business's Facebook page doesn't step on other people's toes. A very good and powerful example are religious beliefs. Taunting or mocking a particular religion's principles or values may generate significant amount of "buzz" for your business and make it popular but such buzz and popularity aren't the ones the can help you in your social media marketing campaigns. In fact, that kind of buzz and popularity will actually harm your Facebook campaigns and your business as a whole. So if you would like to enjoy long term social media marketing success on Facebook or business longevity, it's best to be cognizant about your target audiences' dearly held values and beliefs and be respectful about them, especially if you don't know.

If you make your posts genuinely engaging, interesting and respectful of other people's opinions, rights and values, it's almost sure to say that you will enjoy a much wider audience base that will appreciate and follow your business's Facebook page. As such, you'll enjoy a bigger base on which you can run your social media or Facebook marketing campaigns, as well as a higher possibility for effective and meaningful engagements.

Consider all the people who have experience being bullied on the Internet through discriminatory and bigoted posts on Facebook. The bullies weren't just wrong, their awful posts also backfired strongly on them. At some point, some of those bullies were forced to close their Facebook accounts for fear of retribution and retaliation. And I suppose you don't want that to happen to your business right? So always remember, respect brings respect and fosters loyalty. And when it comes to social media campaign success, those two are very important.

Sift The Important from The Trivial

Another way to successfully conduct your Facebook marketing campaigns is to closely follow the Facebook pages of your business's regulatory authorities, competitors, other similar businesses and your most important followers. With the overwhelming amount of posts from other people that you see on your news feeds on a daily basis, it can be quite difficult to determine which is important or relevant and which is not. This runs the risk of you missing out some important posts that can make or break your very own Facebook marketing campaigns.

In order to address this, its best if you learn to skim through those that are not important and focus only on the important ones. Keep in mind that the most important posts will most possibly come from your business's regulatory authorities, competitors, industry associations and your most significant followers.

Pareto Principle

For all businesses, the Pareto principle, or the 80/20 rule, is critical for success. As a refresher, the Pareto principle states that 80% of results typically come from only 20% of the resources. Applying it to your Facebook marketing campaigns, 80% of your businesses will come from only 20% of your followers. As such, it is absolutely crucial that you know and understand the audience that makes up your 20%.

You need to identify them by looking at their buying habits as well as how engaging they are. A good engagement indicator of a member of the 20% audience is someone who regularly comments or likes your posts. After determining that they are part of the 20%, you will need to really stalk them by looking at their Facebook pages for regular updates so that you will know what's up with them, when very important dates are coming up so you can celebrate with them through greetings, get in touch with them regularly, and continue to stay on their good sides.

Facebook Contests

I can think of no better way to make people enthusiastic then by announcing a contest. As such, I highly suggest that you call regular Facebook contests for your followers, customers and prospects and make sure they know what they will get if they win. Post pictures of the prizes that you plan to give away. Provided very clear instructions concerning the contest to minimize any risk of controversies or disputes.

Certain keywords get people more interested compared to others. These include words such as "grand prize", "winner" and "contest", among others. Ensure that these words can be easily seen so that people will be drawn to your business's Facebook page.

One widely held belief in social networking circles is that emotions are very contagious. as such, it's important to get your self excited and post content that's also exciting for your customers on your business's Facebook page so that they too will become excited. When that happens, the mood will spread like a virus. It can also make more people like your page.

Calls to Action

Calls to action are very important for your Facebook marketing campaign success. You need to tell people exactly what you want them or expect them to do on your Facebook page. A good way of doing this is to literally spell everything out for your followers. Clearly express what is it that you expect or want them to do and embed a link on it. Truth is, most people in social media need to be guided and by doing so, you and your business may experience great benefits from it.

Variety

It's been said that the spice of life is variety. Similarly, the spice of your Facebook page is variety. Make your business's Facebook page very interesting by mixing up its look every so often. You can also post any random pictures, videos or articles that you may have enjoyed recently. What it's true that you need to post mostly stuff that is related to your business and focused on your customers, it isn't a mortal sin to post something that you yourself like or became interested in once in awhile. The key here is doing this once in awhile only for variety. If you do it so often, it may become a distraction and draw people's attention away from your business.

Competition

Checking out or spying on your competition is another way for you to maximize your Facebook marketing campaigns and enjoy best results. There's nothing wrong with doing that for as long as you don't infringe on your competitors' intellectual property rights, which means you shouldn't steal their copyrighted material.

In checking out the Facebook pages of your competitors as well as those of other similar businesses, you'll be able to see the strategies and tactics that may or may not work for your very own business' Facebook page. One good example of this is using what is known as infographics. Do your competitors mostly post infographics or videos? Do your competitors utilize welcome tabs? How frequently does your competitor post and what's the nature of such?

The Power of Images

Using more visual search images in your business's Facebook page can do wonders for its Facebook marketing campaigns.

There's a reason why people with very good or powerful memories are referred to as those with photographic memory. And that reason is the tendency of our minds to think better in pictures. Its the same principle that underlies the saying that a picture paints a thousand words. Pictures or images give you much more detailed information per measure of space, be it online or on paper.

And more than just quotable quotes and sayings, science has shown that images are very powerful communications tools. In some studies, it was discovered that on average, posts that feature images or pictures are able to get as much as 39% more engagement compared to imageless or picture less posts. What's even more impressive? Posts with images or pictures comprise up to 93% of those that are considered as most interactive or engaging.

And finally, post with pictures or images have a 53% higher chance will be getting more likes, have more-than-double the chance of receiving comments and normally get clicked on 84% more compared to posts with no images or pictures.

Timing

One of the best ways to improve your posts' chances of being seen by your prospects or customers is to post status updates or content at optimal times during the day or night. Normally, it's good to post updates and content at the end of the week but specifically, studies have shown that Facebook posts and updates between Thursdays and Fridays normally enjoy up to 18% more interaction compared to posts and status updates during the rest of the week. But if you like more specificity, HubSpot has found that the best time to post on Facebook is between 1 o'clock p.m. to 4 o'clock p.m. In particular, the best or the peak time is at 3 p.m. on Wednesdays.

Now if for some reason you don't want your posts and status updates to be discovered by your business's followers on Facebook, you can also post those between 8 in the evening until 8 in the morning, aside from not posting them at all.

Engaging

Beginning your posts by asking questions is a good yet simple way to interact with your current and potential customers. In particular, posts that start by asking questions normally receive as much as twice the comments compared to those that don't use questions at all. Good questions to ask in your posts are those whose answers are short and limited such as your favorite color, favorite football player or questions that are merely answerable by a yes or a no.

Holding contests on your business's Facebook page is another excellent way to increase engagement. It has been estimated that up to 35% of Facebook fans like to join contests. You can spark your followers' creativities and imaginations when you hold contests that give away interesting and relevant stuff. If you're relatively new to running Facebook contests or simply don't want the burden of doing it yourself, you can use WildFire, Strutta, WooBox, ShortStack, Votigo or North Social.

70/20/10

This isn't a variation of the Pareto principle we discussed earlier. This ratio is very specific for posting on social media platforms such as Facebook. Adhering to this ratio can help you post content with optimal interaction and engagement from your followers.

The 70 / 20 / 10 rule states that:

- Up to 70% of your posts must be those that are able to bring up the value of your business, including its brand recognition;

- Up to 20% of your posts must be shared content, i.e. from outside sources that are properly acknowledged of course; and

- Up to 10% of your posts must promote your most recently released products, latest services offered or upcoming events.

When your Facebook posts are broken down this way, you can maximize your promotional posts while continuing to effectively engage your followers with content that's both informative and enjoyable.

The Power of Feelings

Being emotional is usually considered not good. But in terms of Facebook posts, emotions can be your campaigns' most powerful tool. It goes without saying of course that if you're going to harness the power of emotions, you will need to do it wisely and within reason.

Why should you harness the power of emotions for your Facebook marketing campaigns? Scientific studies have shown that emotions can be very contagious when it comes to posting on Facebook. Practically speaking, posts that are generally happy and cheerful can make people happy and cheerful as well. And on the other hand, boasts that have an angry disposition can be expected to make people angry and reactive as well.

So if you want to maximize the effect of your posts on your followers, why not make full use of emoticons? Many times, using emoticons on posts can result in 33% more "shares" then emoticon-less posts. Emoticons have also been shown in studies to increase your "likes" and "comments" rates by as much as 57% and 33%, respectively.

Focused Advertisements

No doubt about it, you can make full use of all of the free features and services on Facebook to market your business on social media. However, those services and features can only bring you so far, given your business's page features high quality and unique content.

But if you put a little money aside as budget for conducting focused advertising campaigns on Facebook, your chances of engaging your prospects and customers effectively can significantly increase. Remember when we talked about demographics earlier? You can use such valuable information for targeting your ideal market only with paid advertisements.

Insights

Insights is a very useful but often neglected tool for running successful Facebook marketing campaigns. Insights provide information or data on the people that have interacted with or have been engaged by your business's Facebook page. With insights, you can also look at the demographics of your followers as well as just how interactive or engaging your posts are. Such information can help you fine tune your future Facebook ads to make them more engaging, interesting and effective.

Chapter 4:

Master Twitter

The next best social media platform that you should master is Twitter. Twitter is the second most visited social networking site in the world and has millions of users worldwide.

Here is how you can get started with it.

Set Up

The first step is to visit the homeapge, and get started with creating a profile for your company. There, you have to personify your company and fill in all the fields correctly. Try to use a sensible handle and it is best that you mention your company name as the handle. The idea is to make it as easy for your audience as possible to reach you and access your twitter account and website. You have to make your profile as interesting and engaging as possible if you wish to get more and more people to like and follow you.

Having a strong bio that's very descriptive is beneficial for both your business and its prospects. Let's begin with your business. A great bio for your business in its Twitter account is a very good way to introduce itself to many people and grow its organic list of followers.

Now let's turn to your business's prospects. A very well written bio on Twitter can help your prospects who are reading through your profile and who are thinking of whether or not to follow it to know exactly what is it they can expect from following your business on Twitter. And believe me, that is doing them a very big favor.

As you can see, a very well written bio on Twitter is beneficial for both your business and its prospects. Talk about a win-win situation!

If you want to make your bio a very good one, you'll have to make it more descriptive rather than being more creative. Why? Because creative stuff is like beauty - it's in the eyes of the beholder. Simply put, making your business's Twitter bio more creative runs the risk of it being perceived in different ways because creative stuff is highly subjective. Think of it this way, what you consider to be creatively cool may, to another person, look creatively pathetic. So if you focus on making your business's Twitter bio more creative than descriptive, you run the very high risk of driving prospects away.

Since I mentioned that a strong bio with a good amount of descriptions is important, what do I mean by that? It means your bio must include a professional description of your business. Your bio must also have at least one word that is not boring. Your business's Twitter bio must also have a description of its niche or market. The Twitter bio must also divulge at least one of your business's noteworthy accomplishments or your product or service's excellent characteristics or features. And finally, the bio must also include one interesting piece of information about your business, product or service.

Check

The next step is to check if everything that you have added in is correct and in its place. You should share the link with a few friends and family members or staff members and ask them to suggest any changes that you might have to make to make the page look any better.

Tweeting

You need to start conversations and engage people if you'd like your business to gain more Twitter followers and enjoy the social media marketing benefits it offers. Composing your business's tweets yourself is one of the best ways to produce tweets that are very engaging. Yes – you. Not computer programs, not apps and certainly not others. Just you.

Or if you have already hired people to manage your Twitter content, just make sure that they know what they're doing. Anyway, you can monitor your business's Twitter page and see if the people you hired to manage it are actually doing a good job of engaging prospects and followers alike.

For optimal twitter marketing, you have to again keep it really interesting for your audience. Many social media marketers have the habit of just tweeting news article snippets or quotes. Such tweets, however, may no longer be enough to get people talking about your business. So you'll need to tweet things that are interesting so that people will engage you by responding to it.

It'll be good to mix your tweets up. Tweet some quotes, tweet some news article snippets, and tweet some your own stuff. The bottom line is you want to get your followers talking about your business and make them even more interested in your business's tweets. Just remember to focus on your audience and do your best to appeal to as many of them as possible.

It would do you well to tweet more than just text. If you always limit your tweets to text messages, your business's followers and prospects may find them boring and over time lose interest in them. It's best if you can mix in some videos and pictures if you want to make it more fun and engaging for your business's followers and prospects. You can include links to your Instagram posts to help people see the pictures that you are sharing on Twitter. You can make use of its vine feature to create and upload short videos too.

Things may get a little too overwhelming sometimes, with all the multitude of people that are on Twitter. And if you come to that point where you feel there's just too much going on, you can simply "mute" specific people that you follow. This will help you reduce the number of relatively useless and irrelevant tweets that contribute to a clogged feeds section on your business's Twitter page. And don't worry because when you mute them, you can always un-mute them at some other time. It's not as if you're unfollowing them. You're just merely silencing some of them so that you can reduce your Twitter noise or overload.

Interesting tweets are such that are able to interact or engage with people. Always remember that for optimal Twitter marketing campaigns, you'll need to think of engaging tweets for your business's followers and prospects. You will also need – from time to time – to evaluate the quality of your business's Tweets based on the amount of engagement or interactions they generate with followers and prospects. In short, don't make the tweets mostly about your business. Make it mostly about your followers and prospects.

A very good guiding principle for coming up with engaging and interactive tweets is Dale Carnegie's classic bestselling book How to Win Friends and Influence People. Among other things, Carnegie teaches his readers key principles on how to make people like them. These include:

- Being sincerely and genuinely interested in others.
- Remember a person's name because to that person, his or her name is the most important sound and the sweetest name on earth.
- Listen to other people well and encourage them to speak more about themselves.
- When you talk to other people, talk within the context of that other person's interests.
- Sincerely make other people feel important.

What is the stark similarity among all of the principles I just enumerated? That's right, it's all about the other person and not about you. This is because it is ingrained in our nature to like people for make us feel interesting, special and important. Possibly the only principal in Carnegie's book that isn't about the other person is smiling.

They should understand all your schemes and the products and services that you have on offer for them. This step will be a little different from Facebook as the descriptions that you add to each post will be limited to 140 characters. It is best that you draw out the gist of the topic and then post the tweets. It is not such a good idea to post a string of tweets on the same product or promotion as your audience will begin to lose interest in.

When re-tweeting or sharing other people's content, be sure to "hat tip" or HT the originator or author by mentioning them through an @-mention so that they'll be given the credit that they deserve. This is another good way of engaging other people directly and building up your business's Twitter followers and connections.

And speaking of @mentions, using them the wrong way will just neutralize whatever leverage you hope to get by keeping those @-mentions invisible or away from sight of the followers of the people you've engaged via such @-mentions. This is because using the @ of the start of your tweets tells Twitter that such a tweet is intended for the private eyes of the person mentioned and as such, their respective followers may not be able to view it in their individual streams. The only people who will be able to read those tweets are you as the tweeter, the person who was @-mentioned and your common followers.

Another thing that you'll need to master if you want to experience success in your Twitter marketing efforts is consistency. Remember that the Batman-and-Robin of successful Twitter campaigns is consistency and frequency. But unless all you do for a living is to Tweet all day, this can be quite a challenge for you. Running your business, spending time with your family and even times out with friends may get in the way of you being able to consistently and frequently tweet using with business's Twitter account.

What are you to do? Simple – Buffer!

By buffering I'm not referring to what happens whenever you watch a YouTube video using a very slow internet service. What I'm talking about here is using a social media management tool that is called Buffer. You can use Buffer and other tools like that to prepare some of your coolest and most interesting contents to tweet or re-tweet, line them up like teenage outside a Justin Bieber concert waiting to get inside the venue, and instruct Buffer when to release such tweets in your desired order. You can exercise as much or as little control over your tweeting process using Buffer and others like it. Another good thing about using such tools is the ability to review how your scheduled tweets performed using statistics that are recorded.

With Buffer, you can easily set aside one or two hours tops just to prepare your incredible tweet list and tell Buffer to execute their launches for you so you can actually live life.

It's actually ok - and even productive – to tweet some of your posts quite a number of times, contrary to what others think. However, that doesn't mean you can do it with impunity. You can only do it for content that's actually very popular and interesting. When you re-tweet your posts that have content that you and your business' followers love, you can increase the odds of enjoying higher traffic and be able to reach prospects in several different time zones.

Finally, most twitters don't know this but there is actually an ideal time for tweeting. What this means is there are certain times of the day when your tweets are more likely to be discovered by your followers and prospects compared to other times of the day. A survey revealed that the optimal time to tweet your content is between 1 p.m. to 3 p.m. This is because most people will be online at those times or those who are offline will be reviewing tweets posted during this time at later parts of the day. So if you'd like to make the most out of your Twitter marketing campaign, do your best to tweet within this time frame.

Building A Following

The next step is for you to send invites to people to follow you. You should also add a *Follow us on twitter* button on your website in order to get people to follow you. Try placing this button next to the *Facebook page* button on your website and write "and also" in between so that you ask them to like you at both the place.

But a better way of building up your list of followers is to get them to want to follow you and not just ask or invite them to do so.

Responding to or acknowledging people who tweet or comment on your business on Twitter is an excellent and inorganic way to expand your connections by engaging people. That is unless your business is already that famous and has multitudes of followers, of course. And even such businesses can still benefit from acknowledging their followers and tweeters.

Whenever you're mentioned in other people's tweets, when they re-tweet your posts, and tags your tweet as a "favorite", they're actually sending your business an unconscious and subtle message that they want to engage or connect with your business. And that is great because then, there are very good opportunities to interact and engage with them and increase your chances for a successful Twitter marketing campaign.

Keep in mind that unless you are a very famous brand like Samsung, Nike or the Golden State Warriors, it is very important that you acknowledge your business's followers' tweets and mentions by at least thanking them. While it's true that not all of them will consider that significant, some of them may actually re-tweet your acknowledgments or responses, which can be good opportunities to promote your business on Twitter, gain more followers and hopefully, customers.

Following Others

You should follow people and businesses that you think will leverage your own company's account. You have to identify these key people and start following them. You should also re-tweet what they are tweeting in order to draw your own audience's attention to it and remain in the good books of those whose tweets you are re-tweeting.

As the list of people you follow grow together with that of your followers, things may start to become chaotic. You can minimize or eliminate that risk by organizing the list of people you follow on Twitter. If you don't organize it at all, expect many random sponsored content as well as many various avatars in your Twitter feeds coming from re-tweets. Yes, it can be quiet interesting and enjoyable at first but later on, it will result in information overload. That's why having an organized Twitter list is important.

By organizing the people or businesses you follow on Twitter, you can focus exclusively on original, useful and relevant content coming from a select number of people, groups or businesses that you follow without any sponsored content. When you organize your Twitter list well, you'll be able to transform your business's Twitter account into a minimalist one that's relatively easy to manage.

Another challenge for many people on Twitter in terms of following others is reading the tweets of people they follow. For many social media marketers, especially the beginners, the fine line between what's needed and excessive is often blurred or non-existent. As such, they tend to carry a very heavy burden, knowingly or unknowingly, by trying to read every tweet or as many tweets as they possibly can everyday. Before, I used to do that but eventually I learned through the hard and unproductive way that doing so wasn't worth it.

If you'd like to get a clearer picture of what I'm trying to say, consider this. The average person tweets 22 times daily. So if you're following around 100 people on Twitter, this means that on average, you're reading up to 2,200 tweets daily. Yes, you read that right. Daily.

Now consider if you're a really big Twitter fan and follow quite a number of fascinating people, say about 200 of them. Given that the average person tweets about 22 times in a day, expected to read around 4,400 tweets every day.

Do you see it clearly now? Can you imagine how much time you'll need to spend just to go through tweets if you're the type of person who feels obligated to read all of the tweets of the people you follow with your business's' Twitter account? And that's not including your personal account!

The good news is you're not obligated to do so whether socially, legally or from a business point of view. And just in case you really deem it to be necessary and if you have the budget for it, you can consider the next step.

Twitter Team

Just like with Facebook pages, you have to employ a team that is specifically dedicated towards maintaining your twitter account. You have to find those that are adept at knowing what the latest topics are and can easily update the page. Here, they might have to engage in a little trial and error to find the best time when they can update the tweets. They have to post the promotions during different times of the day and find the best time when most traffic occurs on the website.

But in case you're still under budget and can't afford to hire a team just yet, have no fear. There are ways to help you manage your business's Twitter account while working up the budget for it.

The Power of Images

It is important for people to look at the products that you are promoting and for that, you have to upload high quality pictures. You can also place a link to your Instagram account in order to inform people about the products. You can also add in links to your YouTube account in order to get people to see the videos of your products.

Hashtags

Learn to use the hashtags in order to participate in any of the trending topics. You have to know when to use the hashtag to use to your advantage. As you know, the hashtag needs to be used to signify a topic on which you are voicing your opinion. Now say for example you are a T-shirt company and wish to sell your Christmas T-shirts. You have to promote the products by ending your posts with a #merrychristmas or #christmasgoodies or whatever the trending topic is. This will help you remain on top of all the different search engines and can in fact start a viral post if you have enough people reposting it for you.

Collaborations

You have to try and have tie-ups with influencers. These are companies or people that have a lot of followers. If you get them to promote your company then it will only work to your advantage. Even if they are charging you a small sum for it, it will be well worth the effort as you can considerably increase the number of followers that you have. Besides, it will be a one-time payment but the audience that you will get to garner will be for a lifetime.

Quick to The Draw

Try to be a news breaker. It does not really matter what the news is about. Everybody likes to re-tweet something that they think they saw first. So if you quickly tweet something that you know will attract an audience then you can successfully get many to re-tweet you and garner an audience for your company and products.

Reciprocity

You can follow in the steps of many of the big brands out there and immediately follow the people that are following you. That way, you can impress them and get them to be your loyal customers for a long time.

Chapter 5:

Master Instagram

Instagram is a social media platform that allows you to share your pictures with the rest of the world. Let us look at the different things that you can do to increase your Instagram following. Instagram has over 2 million users, which makes it a great platform for you.

Your Profile

Start by visiting this the homepage. There you should set up your profile. Again, use your company's name in the account.

First, you will need to provide a good profile picture. So what makes for a good profile picture? It is one that is very interesting and that can immediately grab a person's attention. It should also be clear enough for people to be able to identify your brand's logo. And if it's something that you have clicked, don't forget to water mark it. Avoid using pictures that are silly or pathetic that you can just download from the Internet. Keep your profile picture fabulously genuine.

Links are good tools to maximize your business' Instagram effect on your overall social media marketing campaign. You can only put a clickable link in one place on Instagram - your business's profile. Make sure that you use this link well so that you can maximize its benefits.

You can change your link as often you feel like doing it. These links can be used to drive people to your business or product's home page, blog page, landing page, custom landing page and contest pages. Just make sure that you use a trackable link creator create such links so that you'll be able to access relevant information about the amount of traffic generated through it later on. The ability to access such information can help you objectively and realistically evaluate the traffic generated by Instagram for the websites mentioned and by that, you can get insights on whether or not what you're currently doing is working or not.

Even if you can't include hyperlinks in your posts on Instagram, you can do so within your posts' descriptions. Your followers can simply copy paste those links on their web browsers in order to go the the website for which the links were posted for.

If you want to be able to reach a wider audience to promote your business profile, you can do so by riding on the success of famous Instagrammers. They are referred to as "influencers" who can help promote your pictures and have it noticed by their followers when they either review it or share it on their own Instagram accounts, thereby giving your posts access to a much broader audience.

Pictures

Start adding the pictures. They need to be high quality pictures that have been taken with a good quality camera. Quality refers to Different aspects of an image such as its depth, light settings, and camera angles, among others. These can greatly influence the quality of the picture that you will take an upload on Instagram. The pictures should speak a thousand words. You have to be creative about it. Just keep in mind that generally speaking, the world isn't forgiving and any mistakes, however small they may be, can be highlighted and blown out of proportion.

If your business is selling physical goods, one excellent way to optimize your business's Instagram account is by taking pictures of your products or services as they are being used or availed of in ordinary, everyday settings. Doing this can give your shots a lively and dynamic look.

But if your business is about intangible products, you can take pictures that present the situations before and after patronizing or using your intangible products or services. To ensure people get to see the important details that can help convince them to patronize your products or services like promos, price and sales, it would be better if you overlay your pictures using texts.

On Instagram, it's your photographs that will be doing all the sales talk for your business. As such, you will need to be cognizant about how you take pictures and not just on what pictures to take. With the right angles and perspectives, even the seemingly simplest or lifeless objects can come alive and look very interesting and attractive. You can even use filters to make your subjects even more interesting. In order to win people's interest, attention and approval, make sure that you take pictures of subjects that have lively and vivid colors for uploading to your Instagram account.

And speaking of filters, while it's good to use them once in a while, avoid being overly generous in using them. Using too many filters in changing of the original too often or too much can make your pictures look weird in Turkey. Use filters only to enhance an aspect of your subject and not to hide things that can be significant in people's decision whether or not to follow your business on Instagram or to patronize its products or services. One good way of using filters is to post a series of the same pictures but with different filters or affect used for each. This can help make your businesses followers curious and come back for more.

If you think you're not that good a photographer yet, no worries. You can always get a professional photographer to come and take high-resolution pictures for you and then upload them. You have to make use of all the different features that Instagram has on offer for the users to use and enhance their pictures. This enhancement will make your pictures look good.

If you want to post a picture of another picture, the best way to go about it is to scan it first then do some editing to make it look good. Keep in mind that whether it's on Instagram or other platforms, no one wants to see poor quality images and if you put such images on your business's Instagram account, not only will people ignore it but they may also leave with a bad impression of your business as a whole. Therefore, make sure the pictures that you upload, whether they're actual shots or scanned copies, are crisp and clear.

On Instagram, captions carry a lot of value. This is because regardless if your posts have little or so many comments, captions are always visible. Because of that, it's absolutely important that you never forget to fill up this part of your Instagram posts.

You can use cleverly-written captions as subtly powerful calls to action. In the case of Nordstrom, it used a caption in one of its Instagram post as a call of action to followers – particularly for them to like that particular post so that the prices of several featured products will be discounted. Without going into much details, let's just say that their clever use of captions caused a highly desirable action from the masses.

Instagram has a very unique feature that people often take for granted. This feature is called square - cropped imaging. Unlike most cameras let's have images shaped rectangular early, you'll notice that Instagram images are perfectly squared, which looks like old photos taken from Polaroid cameras.

Why do you need to keep this in mind? It's because when you take pictures for uploading on Instagram using a regular digital camera, you will need to crop the image later on just so it will fit on Instagram's square frame. If you don't take this into consideration or if this slips from your mind you may inadvertently maximize your camera's entire rectangular window only to be frustrated later when you discover that you won't be able to show everything on Instagram and that you'll need to crop such beautifully taken pictures.

Consistency is key to winning people over – including their patronage of your business – on Instagram. As such, you'll need to regularly post on Instagram. If you don't, people may not notice your business and its products.

And don't allow your account to become defunct because it will reflect badly on your business or brand, particularly on its perceived enthusiasm. So even if it's just one picture, do your best to upload a new one every week. Don't forget to keep your pictures interesting and don't just post any picture just so you can say that you're consistent in doing it. A crappy upload is no better than a no-upload. And speaking of crappy pictures, avoid repeating or reposting a picture that you have already posted before, no matter how beautiful it may be. It may not look much but believe me, that can be a big turnoff to many of your business's Instagram followers.

Building Up Your Following

You should invite people to follow you and can add your Instagram links to all your different social networking sites. You have to link all of your sites for easy access. Don't think just because you have a lot of followers on your Facebook page, you will end up having a lot of followers on your Instagram page. Unless you tell them that you are present there you will not be able to get them to come over.

Handling Comments

Remember our emphasis on being interactive or engaging in social media marketing? When it comes to Instagram, it can even be more important. As such, do your best to acknowledge or reply to all the comments that your posts receive. By responding to comments, people will be more inclined to go back to your Instagram account and view more of your pictures. Even the act of simply sharing an emoticon needs to be done consistently in order to continue interacting or engaging with your Instagram audience.

And speaking of comments, not all of them will be positive. Some of them will unavoidably be criticisms. As such, you will also need to learn the art of ignoring them. Now let me qualify this. By criticisms, I mean critical comments that are obviously meant to put you down and not help you make things better. Constructive criticisms are those given with the intention of helping you improve and for such, all you need to do is take them in stride, thank the person who gave that comment and do your best to address the issues raised by it.

But for critical comments, simply brush them off and not respond to them at all. Trying to convince them otherwise won't work and you will just end up stressed and frustrated or worse, look petty and unprofessional. Now if such people persistently and consistently post critical comments on just about every Instagram post of yours, consider them to be haters and simply block them on Instagram.

Contests

You can organize competitions to keep your audience interested in you. You can ask them to take creative pictures of your products and then share it on your page. Similarly, you have to share any of their pictures and make them the star. Doing so will help you amass a bigger audience.

Collaborations

You can get influencers to promote your pictures. You can ask them to embed your pictures in their blog or repost your pictures in theirs to help you increase your audience base. You might have to pay them a fee, which will be a one-time investment but a great one, as it will help you remain with a big audience.

Calls to Action

You have to give clear call to action for your audiences to visit your website and buy the products. If you are not clear about it then your audience will not know where to go and what to buy. It is your choice to make it obvious for your customers or subtly suggest it to them. But you have to tell them what to do.

Hashtags

You have to use hashtags to promote certain products. It is the same rules that apply to twitter. You have to know when and how to use the different hashtags. You should also remain extremely interactive with your customers. You have to listen to what they have to say and see if they are asking you to do something. Maybe it is to do with the visual style, maybe they want you to edit the pictures out before posting it etc. You have to pay attention to all of it and do as they ask you to.

Maximizing Text Space

Remember that this is a photo-sharing site and so, you will not have too much space to add in lengthy captions and paragraphs. You have to only add in small paragraphs that will fit in the space provided. Make it concise and to the point. You don't want to add in lengthy unnecessary captions that are beside the point.

Links

You should add your Instagram link to your email signature in order to promote it and get others to like your page.

Timing

You have to choose a specific time when you will post your pictures. If you want you can tell your audience that you will be posting next at a said time in order to get them to visit you at the same time. But remember to not over do anything. You cannot keep posting it every now and then or several times a day. Many will miss out on older pictures if you do so. Go about it as a consistent yet calculated speed.

Insta-Videos

You can also optimize the use of your business's Instagram account by utilizing its video capabilities. You can shoot 15-second videos to help your business tell stories that are visually moving and entertaining, which can help promote your business or your brand even more. As with photographic posts, you can edit, filter and utilize them together with other apps like Facebook.

So if pictures paint a thousand words, remember that videos can paint much more than that.

Chapter 6:

Master YouTube

As I mentioned earlier, if pictures can paint a thousand words, videos can paint so much more. As such, wouldn't it be great to be able to reach many more people with your business or brand's videos via the Internet? I believe that at some point in your life, you have already used YouTube or at the very least, have heard of it. In case you still haven't, it's time to learn about and try this wonderfully entertaining social media platform.

YouTube is a great platform that you can use to promote your company, products and services. In this chapter, we will look at the different things that you can do to attract a large audience to your YouTube page.

Set Up

Start by visiting this <u>Page</u>. Here, you should start creating your YouTube channel. Fill in all the fields and the name of your channel should be your company's name. Try to keep it simple and avoid using any special characters, as that will confuse your audience. Write the name of your company clearly. You have to use a clear and attractive thumbnail image to represent your channel. You can use your company's logo if you like, as it will make it easy for people to identify your channel. You can also add in something interesting like a fun picture that will immediately attract people to your channel.

Upon setting up a YouTube channel, you are immediately given a URL that's completely weird and random. If you don't like the URL that's given to you and want it changed, you can ask you to for a customized one. Although most people don't think it's a big deal, it can be an added advantage if you're given a unique URL that can easily remind people of your business or your brand. And after choosing your URL, you need to take permission for that URL. Where there is not such permission will be granted depends solely on the YouTube deities. in fact, you may have to wait for as long as a month or two for such permission to be given. But believe me, the long wait is worth it.

Even before setting up your YouTube channel for your business or your brand, you'd be better off to conceptualize your YouTube channel rather than just being unintentional with it. Example, you can focus your channel the specific kinds of videos like dance videos, fitness videos or music videos, among others. This will be entirely dependent on your business or brand and how you would like to promote it.

Even when it comes to you subscribing to other channels, it would be better if you choose deliberately instead of being random. But using the best ones through deliberate selection and subscribing to them, you make much better use of your business or brand YouTube management time by avoiding excessive and unnecessary videos to go through.

One of the things you can do at the onset of setting up your channel that can also bring in extra income is AdSense. AdSense can help you generate income from your videos' subscriptions and views. It's neither complicated nor hard to use AdSense. All you need to do is sign up using your Google account and YouTube will pay you after the amount of money your videos' views and subscriptions generate increases an earlier set amount. Just ensure that you are on top of it so that you can be paid on time. Typically, YouTube pays for monetized videos every 15th of the following month.

Oh, and as you stay on top of it, make sure you don't get carried away with watching your videos nor be tempted to view your own videos over and over again just to earn extra income for your business off it. They'll know if you're doing that via your IP address and as such, your views won't be counted for monetization.

Branding

Branding isn't just for big and established businesses that have billions of dollars in advertising or branding budget. Even ordinary entrepreneurs like you can do that on YouTube. The cool thing is that it doesn't take a lot of money to put up your own brand or business's channel or to partner together with YouTube.

If you're interested in branding your business or products on YouTube here are some useful tips:

- You can customize your business or brand's YouTube channel's backgrounds and make use of colors that makes accentuate its image. As with other social media marketing platforms, your business or brand can enjoy more credibility and trust when its profile appears to be more branded and customized. While you can use available default designs and templates, doing so will give your business or brand a rather generic feel and may even harm its image on YouTube.

- When you choose Player View for your YouTube channel's layout, you'll be able to upload videos on YouTube that are automatically played. This isn't possible if you choose Grid View for the layout instead. And because your videos' total number of plays can significantly affect its discoverability, choosing Play View to play videos automatically can significantly help your channel's videos become more discoverable.

- While you have the option to make any kind of playlist for your business or brand's YouTube channel, you'd be better off limiting that playlist to the most popular ones or to those videos that are able to best present your business or brand to the viewing public. Better a few great videos than many, average or ordinary ones.

Linking to Your Other Social Media Accounts

You should then link any of your other social media sites to YouTube. You have to try and link everything so that your audience finds it easy to connect with you everywhere. You can also add the *Find us on YouTube* button on your website. You must also add the link of your channel to your email signature to get people to click on it and visit your channel.

Videos

Next, start posting your videos on your channel. You have to post high quality videos that clearly show the product to the audience. Remember that it is important for you to shoot the video in such a way that it keeps the audience glued to it and are able to enjoy it. Make use of a high quality camera for the purpose and shoot the products. You have to choose the HD format, as that is what is mostly used these days. Once the videos are up, you should get your friends and families to look at it and give you a feedback for it.

Because of the intense competition on YouTube, your videos have to be as unique as they possibly if you'd want to have a much bigger audience base for your business or brand. One of the ways to make your videos stand out from most others is to focus on sound and video quality. Don't settle for anything less than your best. While you may not be able to compete with professionally produced videos, you can at least try to create very high quality ones.

Apart from quality, you also need to do your best in making your videos as interesting as possible. Just take a look at some of the most popular music videos on YouTube and you'll quickly see just how amazing and awesome they are visually. Again, you don't have to produce your videos in the same manner as these professionally produced ones but you should do your best to make the videos that you upload be as visually appealing as possible. And visually appealing is one of the things that'll help make your videos very interesting.

It pays to look at your videos once they are uploaded. If you don't like something then edit it before your entire audience sees it. Nobody likes long videos that have a lot of dead content in it. Go till the very end before deciding on going live with it. Have a test audience in place and ask them to give their honest opinion. Based on their feedback, you can either make the changes or ask another audience for their opinion on the same.

Particularly for your products and services, avoid uploading simple demo videos. Again, make them as interesting and fun as possible if you'd like to generate enough interest and tension. More than just demo videos, why not do or create a music video or a short film. Keep in mind that you're doing this as part your business or brands social media marketing campaign, the goal of which is to get more people to patronize your business or product. Your videos will only be able to help you do that if they are interesting enough.

While it's highly improbable that you'll be able to create viral video simply because you don't have a dog that plays basketball or a pet cat that cooks its own food, you don't need such wildly exotic pets just to get a lot YouTube views. Again, all you really need is an interesting, high quality and well conceptualized video and you'd be assured or at least of moderate level of YouTube success. Besides, would you actually want to bear the burden of having to manage a video that's viewed by millions and millions of people? I don't think so.

And more than just creating interesting, well-conceptualized and high quality videos, make sure that you come up with videos that actually address your audience's needs and appeals to their interests. So make sure that your videos are helpful, valuable and compelling enough for your existing followers and prospects.

So how can you tell if your planned video is interesting enough? Well, one way to know that he is to ask yourself if you can write a blog about the topic are the concept of your video. If your answer is yes, it's probably a video worth shooting.

You may be wondering what are the things that you need to include in your business or brand YouTube videos. 41, it should include answers to commonly asked questions, sunscreen video captures, instructions on how to do certain things, slideshows, interviews with the really cool experience and the like.

One of the best things that you can include in your YouTube videos, which can help your business or brand win more followers or subscribers, are annotations. In particular, annotations that include calls for action that viewers or audiences can click. These annotations can be found at the top of YouTube videos, which may include links to other videos, playlists and other channels. Annotations can also contain options for subscribing to your brand or business's YouTube channel.

Videos that contain information that's already passé can benefit greatly from annotations. Instead of having to delete the video because of the update information, you can just make a new and updated one, upload it on YouTube, then embed a link via an annotation on the old and outdated video that will redirect viewers to the new one with updated information. And if you're planning to make many how-to videos for Facebook, chances are that you will find annotations to be one of God's greatest gifts to you.

Lastly, you may be walking on thin ice if you plan to upload videos that you don't have the intellectual property rights for. With videos you created yourself, you own it as far as YouTube is concerned. But if you upload a video that was made by someone else, you'll definitely be violating copyright laws.

At such, you'd be better off sticking to uploading original videos and steering clear of those that aren't. In the event that you really need to upload a video that's not yours, just make sure you're able to secure the necessary permission from its owner. And when you have their approval, you can upload it on your business or brand's YouTube channel.

Don't assume that just because your business or brand's YouTube channel is not yet popular, you can get away with uploading videos that aren't yours. Because YouTube has some very efficient robots, no such materials can remain undetected by its radar for too long – even if your channel is still relatively obscure.

You have the choice to upload advertisements, demonstrations, product descriptions, recipes using your products etc. There are just so many different things that you can do on YouTube. You should use some good and soothing background music in your videos to help people have a fun time while viewing your videos. But make sure that you have the rights to use the music lest you get into trouble for it.

You can upload other videos as well, if it is interesting. This can be getting someone else to use your products and add reviews. Similarly, you can capitalize on the power of your fans and get them to submit videos to you, which you can upload on your channel. The idea is to be as unique as possible and keep it interesting.

Playlists

Organize your videos. You can create a playlist so that the videos can play one after another in your channel. That way, your customer does not have to navigate through your videos to watch them. The power of suggestion always works and you will see that your audience has taken a liking to the list of videos that you have put up.

Visibility

You have to fill in the Meta data of all your videos in a way that it helps them rise to the top spot. This extends to all of your social media promotions. The number 1 spot is extremely coveted and you have to strive to reach it. You have to get your description tabs right to make progress.

You have to fill in the Meta data of all your videos in a way that it helps them rise to the top spot. This extends to all of your social media promotions. The number 1 spot is extremely coveted and you have to strive to reach it. You have to get your description tabs right to make progress.

Timing

You have to have a set timing to upload the videos on your channel. Your audience should know to expect the next video on your channel. Once you establish a set pattern, your audience will start visiting you regularly to check out the new videos. The new videos should be much interesting and what your audience wants to watch. It is a good idea for you to upload something that is educative. Better known as edutainment, you have to give your audience something that they can watch and get information from. This information need not only be in regard to your company and products and can be about anything interesting.

Managing Comments

It is important for you to read all the comments that your channel and videos garner. The important thing is to remain as interactive as possible. Don't make the mistake of falling into the trap of some annoying users whose main aim is to annoy you and evoke a reaction from you. You have to develop a thick skin, as you cannot get stuck with the mean comments. You have to focus only on those that are your loyal customers and are contributing towards your company's growth. You have to answer the questions that they will ask and give them information about the products and services.

Collaborations

You have to tie up with some of the other bloggers from time to time. This will help you generate a lot of leads. Look for people who are quite popular in your particular field and get them to guest blog for you and upload videos of it. You can also subscribe to other channels that will help promote your channel. Try to tie up with those that will genuinely bring in a bigger audience for your company. Some might charge you a small fee but that is fine.

Calls to Action

Ensure that you provide your customers with a clear call to action option. Tell them what you want them to do at the end of the video. You can also add in the message in the middle of the video to make it easy for your audience to do what you want them to. Be clear about it and straightforward. There is no point in making it cryptic. You have to tell them to subscribe to you or visit your website and place an order for the product. You have to literally spoon-feed it to them.

Monetizing Your Videos

If you are able to generate a big traffic for your YouTube channel then you can monetize it. You can sign up with AdSense and AdWords and you will get paid for it. AdSense will start playing an ad that will advertise a product relevant to your video. Every time one of your customers clicks on the ad, you will be paid a certain amount. But the rules might be different for some users where they will be paid only if the clicker buys the product after going to the site by clicking on the ad. Once you reach $100, the money will be added to your account that you would have registered.

Chapter 7:

General Tips on Using Social Media for Marketing

Here are a few general tips on the subject for you to follow.

Reaching customers

You have to establish a direct line of control over your customers if you wish to keep them interested in you for a long time. You have send and receive mails and keep track of what they are up to on their social media websites. It should go beyond mere birthday wishes and you should make the effort to congratulate them for their various achievements in life.

The 80/20 analysis

The 80/20 analysis is one where you perform a statistical analysis to see what is working well for you and what is not. For this, you should start by looking at the numbers. The numbers here refer to the number of followers that you have for your company on each of your social networking sites. Write down the numbers individually next to the names of the sites and then total all to see the final number. Now divide each of the numbers individually with the final number to see what they put out. The one that puts out the lowest number here is your best site. That is the site you have to pursue if you wish to make the most of your marketing strategies.

Organize events

It is important for you to organize certain promotional strategies and events to bring together your existing customers and also educate new ones about it. These events should aim at increasing the reach of your social media strategies. The event should be organized at a venue that is close to a majority of your loyal customers. You have to invite them well in advance and tell them what to expect at the event. You can organize food and drink as well. There, you have to ask them to educate their friends and family members about your social media strategies. You have to inform them about any upcoming deals and schemes that will interest them. Before they leave, you should ask them to leave behind a feedback asking about their opinion and also if they will be interested in attending anymore such events.

Social media only offers

You have to offer your customers a social media specific offer. This means that you give them the chance to make the most of their social media presence and be thankful for it. You can give them *social media only* discounts or *social media only* giveaways. Just ensure that it is a big amount so that they remain motivated to keep buying from you. You can also offer them combined benefits by tying up with another company. Those that like your Facebook page or re-tweet one of your tweets can be given a discount coupon for another brand. This will help you expand your customer base in a big way.

Apart from the four different social media sites that were mentioned in this book, there are some others like linked in and Google+ that you should exploit and use to your advantage. The more presence that you have on the Internet, the bigger your audience will get. So try to sign up with all the different platforms and use all of them to your company's advantage.

Chapter 8:

The Dos and Don'ts of Social Media Marketing

I n this chapter, we will look at the different dos and don'ts of social media in this chapter.

Dos of social media marketing

Repurpose

The first do of any social media-marketing scheme is repurposing the content. This will ensure that you save on both time and effort. You must know what to repurpose and how you can use it again. Go through all the content that you have used before and if it is possible for you to repurpose and use it. You have to modify it in a way that makes it look like brand new content and then use it to your advantage. But avoid simply copying and pasting as that will not work well for you.

SEO

You have to know to use SEO to your advantage. You need to look into the different aspects of it and know how to be discovered by many people. You have to know how to use the key words, Meta data, Meta description, the different headlines etc. All of these will count when you wish to reach out to a large audience. If you are not able to do this on your own then you can also take the help of a friend or expert in the field who can do it for you. Don't take this step lightly as it will be extremely important for you to be noticed by people if you wish to increase your sales.

Customer Involvement

You have to have customer involvement if you wish to become a popular brand. You have to get them to promote your brand and make them your brand ambassador. You can promote your products by using their pictures and get them to advertise for you. Your other customers will also be quite impressed with your ways and start to contribute towards it. You have to conduct competitions in order to get the customers to get involved in your business and visit your pages on a regular basis.

Back up plans

You have to always have a backup plan in place. In case something goes wrong, you will have a second plan in place that will spring into action. You should always remain calm and composed no matter how bad a situation goes. Maybe your promotional scheme will completely flop or a certain section will take offence to it and make you take it down. All of this will impact your promotional strategy negatively. So, you have to be prepared for the worst and have a plan in place that will allow you to quickly take action and save the day.

Unique

Remember to always be unique. Don't make the mistake of copying someone or trying to look like something else. You have to adopt a promotional strategy that is unique to you and use it to your advantage. You will see that it is working in your favor and helping you avail many customers. You will also see that you have unique customers and not the same repeat ones. You have to try and re-invent yourself from time to time and give your audience something new to experience.

If you do all of these then rest assured, your company would do well in the social media-marketing field.

Don'ts of social media marketing

Do too much

It is extremely important for you to know where to draw the line. You have to know when and where to stop when it comes to interacting with your audience and getting involved in their life. You cannot do too much and start stalking them or start romancing them. That is not part of your plan and should not be pursued. You have to try and remain as professional as possible and put your promotional ambitions ahead of everything else. You will see that it is possible for you to connect with your audience in a better way if you keep it professional.

Not tap into potential

You have to exploit all the potential that your social media platform offers to you. Explore all of its features and make sure that you are using everything to your advantage. Many times, we will either be less informed or too lazy to exploit a certain feature. That has to be sorted out first and only then can you make the most of what the social media has on offer to you. Take the help of a programmer if you are not sure of how everything can be used.

Repeat the same

Do not make the mistake of repeating the same content on all your different platforms. You have to keep the same tone but present the products in a different way. You have to bear in mind the type of group that exists on the particular platform. You must change the layout of the promotion, the over look and feel of the presentation etc. All of it will go a long way in helping your audience remain interested.

Not understand needs

You have to understand all your customers' needs and cater to them individually. There is a difference between pretending like you care about them and really caring for them. You have to take keen interest in understanding what they want and whether or not they are satisfied with what you have on offer for them. If they are voicing their opinion then encourage them and do, as they want. You have to employ someone to answer all the questions that these people post. Answering them will help you connect with the audience better.

Updating

You have to update your account from time to time. Again, employ someone who will do it for you. You have to choose someone who is in sync with your thinking and know exactly what to say. You should keep your audience up to date with what you have on offer for them. Try to keep it interesting and intriguing so that your audience is forced to look at what you are promoting.

These form the different don'ts of social media marketing that you need to bear in mind if you wish to make the most of it.

Chapter 9:

Social Media Marketing Success Stories

Social media marketing has managed to take the world by storm. Social media giants like Facebook, Twitter, Pinterest and LinkedIn are still the go to social media platforms for most of the digital marketers today. But with the arrival of an increasing number of social networking applications, the digital marketers have an opportunity to look for different ways in which they will be able to expand their brand into different social networking platforms, increase their awareness and also broaden their scope of marketing. Of late it seems like brands are more than willing to turn towards newer social networking sites and are much open to try and experiment with their campaigns by making them more edgy, funny and something that is different from their usual campaigns. Brands are trying to break free of their preconceived image and offer something more to their customers. Some of the companies that have managed to create quite an impression on Instagram are Nordstrom, GoPro and Spotify, some of the unlikely names such as Hilary Clinton and even Taco Bell have managed to garner publicity by making use of Snapchat. In this chapter, let us take a look at some stories that have managed to turn some of their craziest ideas into successful campaigns.

GoPro is an adventure camera manufacturer and they had come up with a really unique marketing campaign on Instagram that featured videos of felines that were skateboarding from Australia. Their campaign had managed to create a lot of buzz for their brand. The reason was that who wouldn't want to watch a video that manages to capture our cute furry friends engaged in human activities? They had managed to make a video go viral by combining the obsession of web viewers with cats doing goofy things with the genuineness of a consumer made video.

Expedia is a travel site and it had come up with a campaign that made use of the social networking platform Instagram. In this campaign Expedia was offering a free trip to South Africa to those who participated in this campaign. The reason why this worked was because they had made use of a social networking vehicle that has got millions of people hooked across the globe. The promise of an exotic holiday is too good to resist for anyone and that's why they had managed to stir up the interest of the masses. For instance, a single photo of Rio de Janeiro that was posted as a part of the campaign had managed to get around 499 likes and more than 500 comments within no time.

TOMS is an e-Commerce vendor and they had set a goal that for every pair of shoes they would manage to sell, they would donate a pair of shoes to the destitute and needy children. This philanthropic campaign of TOMS was showcased on Instagram. TOMS mission was quite noble. Whenever someone would post a picture of their bare feet with the hash tag #withoutshoes, TOMS would donate a pair of shoes to the needy. They wanted to demonstrate the dedication their company had towards the cause they had committed to. They had managed to donate 296,243 pairs of shoes altogether.

Taco Bell has been targeting millennials by making use of a platform that happens to be quite popular in their demographic, Snapchat. Taco Bell had decided to make the most of this platform and it proved to be a smart decision. Taco Bell had launched this campaign for launching new items on their menu and held a contest of uploading snaps to its Snapchat story on the app. They had over 200,000 followers and their snapstories started generating a lot of hype and created quite a buzz.

Nordstrom had decided that they would announce their biggest annual sale by posting about it on Instagram. The company had written a post on Instagram that announced that their biggest sale of the year deserved an equally big message on Instagram. This upscale retailer had managed to create a gigantic art installation that was around 13,398sq. feet high and it also had a 25 feet long dress, this was installed on the roof of their flagship store in Seattle. The related post on Instagram turned out to be a huge success and they had received more than 16,000 likes. The sheer size of the installation showcased in the video represented the size of their annual sale and this managed to grab the attention of all the eager shoppers.

The famous international beauty brand L'Oreal Paris had set up a really elaborate digital hub in New York in their studio. This included eight of the prominent style influencers from the fashion industry and several models as well. The team would focus on recreating the looks that were sported by the celebrities who walked the red carpet during the Golden Globes in the year 2015. Once these videos had been shot they would be turned into GIFs that were shared on various social networking sites such as Tumblr, Twitter and Instagram along with the links to their e-commerce site. These GIFs were of high quality and their innovative idea did help them gain quite a fan following.

Spotify had recently launched a new tool that would scan the playlist of the user and their history as well and inform them of the artists that they had been the early listeners of who had later on become quite popular. Spotify had decided to take this tool to Instagram, a social media platform for its promotion and had started using the hashtag #foundthemfirst. This campaign for the promotion of their online tool was successful because Spotify had decided to keep it simple and didn't unnecessarily over think it. Making the user data available that the company had with it was a really good marketing tactic. Spotify is all about connecting fans to the music that they appreciate and also giving them the inside details of this world.

HP (Hewlett Packard) is not really known for their prowess on social media and the social media campaign strategy that they had designed for the promotion of their HP PavillionX360 convertible laptop had changed the viewer's perception of HP. Their campaign on Vine was titled as #bendtherules. Instead of simply buying social ads, they had made use of paid social influencers on this network.

So, they gave the product they were intending to promote to these social influencers who in turn helped in creating awareness about this product on their Vines. One of the popular Vine videos that had garnered a lot of attention was that made by a popular Viner named Robby Ayala. In the video that was posted, Ayala tries bending the competitor's laptop that resulted in quite a comic video and had managed to get over 250,000 likes and more than 7,000 comments!

Critics have often written off Hillary Clinton as being too cold, calculating and controlled. The PR team behind the Clinton campaign decided to change this general perception of the public by making use of a social networking platform and they chose Snapchat for this purpose. The Snapstories that were posted by Hillary were successful because they managed to humanize her ill-conceived character and made her seem more relatable than all those speeches that she had ever made.

Burberry, the British fashion house in a move to promote their clothing line had decided to make use of Periscope for streaming live their menswear fashion show and they also made use of Snapchat for providing a fleeting photographic update of the event. This helped the iconic brand receive the best digital engagement that they had probably ever received. A month after this the fashion house had decided to stream live their fashion show that was held in Los Angeles on Snapchat and this helped them create over 100 million impressions.

This proves that if you know what your target base is then making use of social media marketing will prove to be really beneficial. If you just know this one thing, then you can make use of the best resources available and create a marketing campaign that is beautifully strategized. Such a well-designed campaign will provide you with amazing results.

Chapter 10:

Ten Things to Be Kept in Mind

Social media marketing seems to be the "in" thing these days, but there are some important things that you need to keep in your mind while you are making use of this in your marketing strategies. You need to come up with a mental checklist that will let you ensure that all the efforts that you are putting into this campaign are not just goal oriented but also capable of being measured, while proving to be successful. In this chapter we will take a look at ten things that you need to keep in your mind while you are developing your social media marketing strategies.

This is not just an additional platform

The broadcast marketers might make you believe that these social networking platforms are nothing more than an additional platform that will let you distribute your message to the masses. Well, this really isn't true. All the different companies and organizations that solely focus on their press releases and also keep talking about themselves will definitely end up as social media road kill. You don't want this, do you? Social media happens to be an important platform where you can engage with your existing customers and attract potential customers, so put in some effort and develop an engaging campaign.

Beware of social media experts

Beware of these so-called social media experts; they are simply sprouting up out of nowhere. Everyone and anyone claims to be a social media expert these days. They all can talk the talk but not many of them can walk the walk. Simply having the knowledge about how to tweet doesn't make anyone an authority on this subject. If any of these so called experts can manage to show that they have had years of productivity, have a good reputation, come with good and genuine references and have managed to produce results that can be quantified. Only such a person can be genuinely called as an expert. So, you needn't worry and you don't need a social media expert, as long as you know what you are doing and how to do it you will be fine. Be wary of anyone who claims to be a social media expert.

Some things never really change

The world of marketing has had a complete makeover; the way it is being approached and looked at has also changed. But there are some things that never really change even after the advent of all the technology. The good ol' rules relating to communication, public relations and marketing still hold good. These old fashioned rules can be thought of as the basic ethics and they will never go out of fashion. It is crucial to know about the dynamics of your target audience, the value that your organization can add to their lives and also your ability to develop and offer such products and services that can satisfy their needs and wants are the areas that you really need to concentrate on. And this information can prove to be very valuable for your marketing campaign. So, do not stray away from these concepts just because you have opted for social media marketing.

Social media marketing isn't restricted to just Facebook, LinkedIn or Twitter

Yes, Facebook, LinkedIn and Twitter are really popular social media sites but these aren't the only social networking platforms. In fact, they just form a tiny fragment of the social media ecosystem. Web forums, email lists, podcasts, social bookmarking sites, different video and photo sharing services and niche online communities all form a part of the world of social media. You need to keep in mind that you haven't restricted your campaign to just one platform. You will need to make an effort and understand the different platforms that your customers are making use of and start getting involved on those platforms to attract more customers.

It's all about creating and maintaining relationships

At the end of the day, marketing is all about creating and building relationships. And social media provides you with all the necessary tools and likely platforms for you to get this job done. But this does not mean that you should ignore the basics things such as one on one personal communication, because this helps in building a good rapport with your customers. You can't build a relationship overnight, it takes a while. It is not just about the technology that we make use of; it's about the human beings.

Don't get carried away

Social media is a powerful tool, but you need to realize that it can help you do a lot of things but not everything. Considerable time and effort should be dedicated towards developing a marketing strategy that would actually work. You will always have to do your bit if you want to achieve your goals. Social media marketing will work only if you have a good marketing strategy. So, put some effort into it.

It is not just confined to one thing

There are different elements that you will need to combine in a harmonious manner if you really want your campaign to work. Social media would be just one element; you will have to ensure that all the other elements of the campaign would also go hand in hand with this. You cannot completely ignore the traditional methods of marketing and should develop a marketing campaign that would strategically involve both the new and the old methods. You should give equal importance to all the other elements as well, instead of just solely depending upon social media.

Building your brand

Some of the organizations these days are slowly moving away from destination websites and are instead shifting their focus on various community building strategies. If you really want to build your brand, then you will have to concentrate on various other things as well. It is a really good move to have established your brand on major platforms such as LinkedIn, Facebook, Twitter or even Instagram, but what you need to remember is that these aren't your only options. You need to understand that your existing customers as well as potential and likely customers might be active on various other platforms and affinity groups that aren't a part of the above mentioned popular platforms. You will need to broaden your horizons.

It isn't just about return on relationship

You should be able to measure your success by making use of a mix of qualitative as well as quantitative metrics. But apart from it, you should also concentrate on different aspects such as brand recognition, reputation of the brand in the market and the public's awareness. But you should also pay equal attention to metrics such as the money that has been raised, any increase in the number of attendees and subscribers, also any increase in the widgets that are sold and so on. All these things will give you an idea about whether or not your business is doing well. You will have to be able to track the changes that have been produced by using social media marketing, if and only when you are able to obtain this data you will be able to know whether or not your social media strategy has worked.

You need to be sociable

You need to remember that at the end of the day there is one thing that is very important. You just need to be sociable, after all social media is all about being sociable. No one and I mean no one would be keen on being associated with someone who is antisocial, not sociable or just keeps broadcasting. You will need to come up with a strategy that will make your brand seem more socially appealing to your potential customers. It is about creating a good reputation for your brand.

Chapter 11:

Social Media Tools

There are various social media tools that you can make use of and it might be overwhelming to decide which of these tools you should make use of. To help solve this problem, in this chapter we will take a look at some of the best social media tools that are really helpful.

Mention: This is a lot like Google alerts, but instead it is designed for social media. Like the name suggests this tool can help you to monitor your web presence on the Internet in an effective fashion. Mention also comes with a few features that allow you to respond to all the mentions made to your brand and it also lets you share any news that you have come across with the rest of the industry.

Buffer: this is an analytical tool that incorporates social media publishing within it. Buffer is a really handy social media tool and this helps in sending your updates across to the titans of social networking platforms such as LinkedIn, Facebook, Twitter, Google+ and so on. It comes with an inbuilt analytical system that not only help s you understand the reason why certain posts seem to be working better than the other posts as well as helping you figure out the best time for posting any particular publication according to the needs of your target audience. The features offered by Buffer do not end here, it also lest you to work collaboratively with your team and keep updating the account with fresh content on a regular basis.

Feedly: This is a content discovery tool and it is really helpful for finding content. It not only helps you find good content but it also lets you share whatever you have found with the right audience without any difficulty. You also get to subscribe to the RSS feed so that you can stay in sync with all the recent updates on various blogs related to the industry as well as upcoming sites. If you happen to be interested in one particular topic, then you can make use of Feedly for tracking content related to your area of interest.

Twitter Counter: The name is self-explanatory. This online tool lets you keep a track of all the changes related to your followers and it also helps you in making a prediction relating to the growth of your followers over a period of time. After a point of time, it gets really hard to keep a track of the growth of your Twitter account. This free service really does come in handy. It can help you to understand the rate of growth of your followers and it makes use of this information to help you decide upon the content that should be sharing with your audience. Not just this but it also helps you decide if there is any older content that you should probably share once again so that they new users also get a chance to view it.

Zapier: This social media tool acts as a platform where you can connect with all the various services that you make use of and it also lets you synchronize them all thereby making your work much simpler. Your team might usually make use of HipChat for communicating with each other, and then you can make use of Zapier to set up an option that would enable automatic notifications in the HipChat chat rooms for any latest updates. You will also be able to connect all the different apps that you use on a single platform and this definitely simplifies your work.

Bottlenose: This comes with a brand new feature that has an inbuilt real time search engine that helps you consolidate all your work from social networking sites and various other groups and then it displays the resultant data in an order or according to algorithmic importance. The end result is the content you obtain that has been arranged according to its importance in descending order. When you have all the information that has been arranged according to your needs then your job does get a whole lot easier and you also get to share the results you have obtained. Another feature is that you can also opt to integrate Buffer with Bottlenose and this allows you to add any additional content as well as resources to it so that you can make use of it on a later date if you don't want to overwhelm your followers with a lot of information.

Followerwonk: You can always make use of an analysis tool like Twitter Counter for instance for keeping a track of changes in the number of your followers, but then again there might be some cases wherein you might have wanted to go little deeper and analyze your followers a little more carefully. In such a case Followerwonk can be used, even this is an analysis tool for Twitter. This tool provides you an analysis of the followers by simply segregating them according to different demographics and this lets you get a better understanding of your followers. Perhaps the noteworthy feature of Followerwonk would be its ability to give you the specific time at which your followers are online and this helps you decide the time at which you should probably tweet to ensure a wider reach to your audience. This tool also helps you know the time at which the people whom you follow are online as well. This really comes in handy when you have been meaning to start a conversation with someone you are following.

Quintly: This really is a very powerful tool and this can be used for obtaining a detailed analytical report of social media and it also helps you keep track of your business on various social media platforms like Facebook, Instagram, Twitter, YouTube, Google+ and so on. Not just this but Quintly also lets you benchmark those features that can help you compare your performance against that of your competitors as well as against that of the industry as a whole. The Quintly dashboard lets you customize the tool according to your convenience.

Chapter 12:

25 Tips to Social Media Marketing Success

Social media marketing might seem very confusing and it is really easy to lose focus while making use of these tools. There are different platforms that you can make use of depending upon your conveniences, but before you zero in on any particular platform you need to realize that each of these platforms has got a definite format and particular rules that you will need to learn about, only then will you be able to use it for your advantage. There will be plenty of time to expand your knowledge of these things later on, but before that you will need to get acquainted with the ones that you have opted to make use of.

If you have decided that social media marketing is indeed the way to go for you, then the 25 tips mentioned in this chapter will definitely prove to be really useful for you.

Make use of only those platforms that are suitable for your business needs:

Just because you have decided to make use of social media in your marketing strategy does not necessarily mean that you will have to make use of every single platform of social media that is available. You need to realize that it takes a considerable amount of time for understanding and learning about every one of these platforms before you can implement your plan. Not even the most seasoned of the marketers will be able to handle more than a few of these accounts at a time, therefore you will need to spend a considerable amount of time and take a closer look at them all, look what are the benefits it would provide for your business and also whether or not it is suitable for your business before you get started. Twitter and Facebook might have the largest number of subscribers and active participants, but this does not necessarily mean that these are the only choices you have got. You can also look at the different smaller platforms; you never know which of these might prove to be the best option for you.

You will have to evaluate everything:

The only way in which you will be able to determine whether or not the efforts that you have put in are working effectively and in the manner that they are supposed to work in can only be done by evaluating the data. There are some social media platforms that have built in tools that will help you do this, if not you can always make use of the third party social media tools to for the purpose of analyzing your data. You can make use of these tools for understanding which part of your content seems to be getting maximum responses or even perhaps the content that isn't able to attract more users and so on. In this manner you will be able to come up with a strategy that is suitable for your business needs.

Make sure that you are posting the right thing at the right time:

It is not just about what you are posting that will have an effect on the number of views that you can generate, but it also depends upon the number of people who are sharing and interacting as well. The timing of your posts is of utmost important, most B2B business just stick to the ordinary working hours for posting, but even with the given uniformity in posting, there will definitely be some days wherein you will be able to get a much better response or vice versa. You will need to put in some considerable efforts and figure out who your target audience are and the likely time they would be online, depending on this you can make a decision.

Start building your connections:

One of the most common mistakes that a lot of social media marketers seem to be making is that they are talking at their audience instead of talking to them. You will need to engage with them on a personal level, interact with them and start building your connections. Your followers would rather associate themselves with a human being than with an automated machine. You can build your connections by asking them to share their thoughts and then responding to what they have shared. If they send you any messages, then you should respond to them as soon as you can and if you start ignoring them, you can forget about any future potential conversions.

Going visual might help:

People are often put off by too much of text; a large block of text wouldn't be able to generate as much interest as a video would. Start making use of images, videos and even info graphs, these are visually more appealing and also contain the information that you want your users to see. When you are making use of any visual content then you will have to make sure that it is appealing, relevant and makes sense as well.

Make each one of your chosen platforms unique:

There are various tools that you can make use of for sharing your content across various platforms, this usually works when you have to share information that is really important, but then if you keep doing the same for every piece of information that you are sharing, then all the platforms that you have chosen will seem to be the same. You need to realize that the people who are following you on one platform are likely to follow you on different platforms; they wouldn't want to keep seeing the same content over and over again. You will have to make each of your accounts different and unique, they should all be equally engaging and appealing, this will help you gain the interest of potential customers.

Make it worthwhile for your followers:

When someone starts following you on any social media platform, they would want to feel appreciated for doing so. To make it worthwhile for your users you can offer rewards for subscribing or even following you, this can be in the form of a small discount or even an entry to a lucky draw for a prize. An incentive will be sufficient motive for an individual to start following and this will definitely grab their attention.

You will need to be a personable person:

Yes, it is true that social media is more relaxed than ordinary marketing strategies of a business but you will still have to maintain some professionalism about the way you do things. You should give out some personal details that will provide a human character to your business but this does not mean that you start giving out your personal opinions about different things on the official page of the business. If you start posting your personal views on topics related to politics or start talking about the latest celebrity gossip, it is very likely that your r followers will stop following you.

You need a social media manager:

This might not seem like a real job to you, but the results of having a proper social media manager will be really helpful in generating outstanding results. Not everyone is really good at managing social media, and if this does not happen to be your area of proficiency then you really should engage the services of someone who can get the job done and is good at it as well. In this manner, if you have someone helping you out with these things, then you can get back to running your business smoothly without having to divide your time and attention towards the marketing campaign. All that is required of you will be supervision.

If something doesn't work, then let go of it:

Not everything you select will work for you; if something doesn't work for you then it is smart to just drop it. It really doesn't matter how much time you have spent analyzing it, the number of team members who are working on it, there will always be that one platform that simply isn't the right fit for your business. Instead of wasting further of your valuable resources on trying to make something that clearly isn't working work, you should just let go of it. There are more opportunities out there that are waiting for you.

Start building relationships with different businesses:

If there happen to be any businesses that are in the same sector or industry that you are in then you can friend them and start following them, provided they aren't your immediate competition. You can always refer customers to each other and share followers and while doing this you might also be able to pick up on some tips. This might really come as a pleasant surprise the amount of things that you can learn from each other. But this does not mean that you start following everyone; show some discretion and chose wisely.

Learn to face the trolls:

As your success and popularity starts increasing the amount of attention that you will receive on social media will also increase. And this means that you should also be prepared to deal with some abuse from other people. If you notice that you have haters then you should be really careful in the manner that you reply to them and you can also bloc them if it starts getting too much. But don't block individuals just because they don't like your company or brand, this does not send the right message across to the other users.

You should always leave your work at work:

Especially when you are starting out, it is really tempting to be there 24/7 and keep a watchful eye over how things are going on, but you need to realize that this really isn't possible. Mobile technology has indeed made it possible for you to check your performance on social media from the comfort of your own home, but it really won't do you any good to keep replying to comments and queries in the middle of the night. The best thing to do is not to download any of the applications that are related to your work on your phone. You shouldn't be carrying your work home even after the business hours. Once the working hours are over you should remember that your work for the day is over as well unless and until it is a really pressing matter.

You don't have to keep on selling:

This might be the reason why you are in business, but it really isn't a good idea to keep selling, this is considered to be a huge sin on social media. Your followers wouldn't want to keep getting bombarded with sales pitches after every ten minutes. You should understand that they aren't on social media to do you a favor; instead they are there to enjoy themselves and to learn probably. All they are looking for is a way in which they will be able to build relationships and social media is for their entertainment. When they want to buy something, they will invariable come and find you, it might be on social media or it might be on your website or any other means.

Make sure that your business profile is complete:

All the social media platforms will provide you with sufficient space in your profile to give some information about yourself. You should make sure that your profile isn't incomplete and also keep updating it regularly. Your followers would definitely want to know about you and leaving blank spaces will not make you more appealing to them. They really wouldn't want to follow someone whose profile isn't even complete.

You should make social media a part of your plan:

Think of social media as a part of your business, an area where you should direct some of your focus. Just like with any other area of business for you, you will need to make sure that you have set suitable goals and objectives to ensure that you will have a business plan to follow. And also when you have set goals you will be able to measure your performance in a much better way.

Make your followers eager to see your updates:

The goal of any good marketing strategy would be to get your followers to be eager to read the content that you post. Your followers should be hanging out to every word that you write, they should be eager to read your next post and they should be enthusiastic about it as well. You would want them to keep checking constantly to see whether or not you have posted anything new. The only way in which you can get your followers this excited about your profile would only be by posting content that is of high quality, it is relevant and interesting to read.

Your content should be easy to share:

Technology has managed to simplify a lot of things for us. You can derive the most of anything by putting in a little bit of effort. And the same is true for social media marketing. But if you really want your content to be shared then you will have to do a little bit of work. You should be able to package your content in a manner that is easy to share and also give people the buttons that would allow them to share the content with their friends or followers through their own social media pages. You should make it really easy for your content to be shared that people will not be able to resist not sharing it. When the content you post keeps getting shared then your potential customers keeps increasing and the conversion rate will also go up.

If you share something, then make sure that you comment on it:

It is not just about clicking the share option that lets you share something or even repost it. If you add a comment to something that you are sharing or reposting it would show that you are sharing it for a particular reason. It would make the content seem like it is worth sharing with others. This will help you build up not just your own expertise but it will also help in building your reputation for being such an expert. This tends to add a lot of value to whatever you are sharing in future. Your personal opinion for sharing when added in the form of a comment makes your followers get to know you better as well.

Keep checking for any grammar or spelling errors:

This is a really important step. You are a professional businessperson and the worst thing that you can possibly do is publishing content that is full of grammatical errors and rookie mistakes. If you think the content you are about to publish is anything short of well written then it will really do you good if you don't publish it. You should check for errors and then proof read it, double check your work and read the final draft once again before you publish it. The content you publish should be well written and it should be of high quality.

You should learn that reply and mention isn't the same thing:

You should get yourself accustomed to the basic features of the different social media networks that you are making use of. For instance, if you start a tweet with a username, then the people to whom this tweet will be visible to would be you, the person whose username you have mentioned and anyone who sis either following you or the other person. If you place the username somewhere in the middle of the post then it will make it visible to people who either follow you or the user. If you are trying to talk about a specific person and you also want everyone to see it then you will just have to put that person's name in the center of the post and not at the beginning itself. But there might be some instances where it makes sense to have the username at the beginning of the post and not somewhere else, then in such a case you should add a full stop after the name. When you do this, such a post will be referred to as a Mention and not as a Reply.

Never post on the hour:

Most of the meetings and tasks as well are scheduled in such a manner that they will start at the top of the hour and when the clock strikes, people would be proceeding on to the next item that is on their agenda. And they will definitely not be looking for any updates on various social media networks. If at all a task or even a meeting finishes early or perhaps it overruns a bit, then in such a case there will be a small window of opportunity where they will be able to check the various updates on their apps. Therefore, it always makes more sense to post your content either just before or just after the hour and not right on it. In this manner you will be able to ensure that more and more people will be able to see what you are posting about.

Interact with those users who share your posts:

If some follower of yours shares a link to your content or perhaps re-tweets or reposts the same message or post that you have published, then in such a case you really should take some time out and thank them. You should always keep such people on your side, these people are carrying out free advertisements for you and you should rightly thank them for doing so. Building a good relationship with a user who is supporting you can go a long way. This would ensure that they would keep on sharing the content that you are posting even in the future. Free publicity and advertising can help in building your brand and thanking someone for that isn't much of an effort.

Familiarize yourself with the platform guidelines:

Each and every platform of social media will have certain guidelines. You should familiarize yourself with these guidelines to understand the form of content that is considered to be acceptable and what isn't acceptable. Basic common sense would dictate the kind of content that is acceptable and the kind of content that isn't acceptable. Also you should keep checking on the terms and conditions and the various guidelines that are given by the platforms. For instance, Facebook keeps updating its guidelines related to things like running competitions and so on and breaching the guidelines issued would result in a penalty or they can suspend or even delete your account from such a platform and this can cause some serious damage to your business.

Ensure that you have included your location in your profile:

Always ensure that you have included your location in your profile on social media networks. This will provide people with the necessary information relating to your business. People would know where your business is located even if your product or service offered happens to be based on the Internet; you will still need to provide the location of your business. Only if they know where you are or where they can find you, only then will they be able to check in, especially on Facebook. This really is of great importance when you have a physical store that people can visit and not adding your location might cost you some potential customers.

Key highlights

The very first thing is to understand what social media marketing refers to and know how to use it. It is extremely important for any company to advertise themselves and their products and services. The best way to do so is by making use of social media platforms. You will see that it is possible for you to reach out to millions just by being part of a social media website.

The next thing to do is undertake the step by step procedure that you must adopt in order to arrive at the desired results. You have to go about it in a set way so that you can save on both time and effort. Start by researching the topic of social media marketing and then move to choosing the best platform for yourself. The next step is for you to set yourself up online and then look for the best promotional strategies that you can use to promote your products and services.

Facebook is one of the best platforms that you can pick to advertise your brand and products. You will see how easily you can reach out to millions. You have to make good use of Facebook *pages* and you will see that it is one of the best platforms to promote your brand and image. Facebook is an easy platform to advertise on and it is even simpler to keep it updated. You will find it quite convenient to tell people about your products and services by making use of Facebook pages.

The next social media platform that you can advertise on is Twitter. Twitter, as you know, is the second most used social media platform in the world. You can keep it small and simple and advertise your brand, products and services. You can easy link all your social media platforms using this medium. You have to post Instagram pics on your twitter account and get people to notice it. You have to know how to use the #s and participate in the on-going viral movement and capitalize upon it.

YouTube is the next social media platform that you can use to advertise your company, products and services. YouTube gives you the chance to make use of videos that you can upload and reach out to customers. YouTube videos have the tendency to go viral, which will make it extremely easy for you to promote yourself. YouTube also has the power to demonstrate your products and services to your customers, which you should use to your advantage.

Instagram allows you to click pictures and upload them. These pictures need to be good quality pictures that you can use to promote your products and services. You have to appeal to as many people as possible if you wish to make the most of the platform. You have to get someone famous to endorse for you as that will get you noticed by a lot of people. Instagram is possibly the best platform for you to advertise and showcase your products as a picture speaks a thousand words.

We looked at the different dos and don'ts of social media marketing that you have to bear in mind. It is important that you understand them carefully and do only as is asked of you to. Over doing something will only hurt you and your company. Put in the right efforts and you will see that it is simple for you to appeal to the right audience and increase your customer base. You can go through the don'ts again and steer clear of them.

Remember that the customer is always king. When you keep them happy you will see that your business is growing in leaps and bounds. So, you have to listen to what they are saying and keep them satisfied. Don't unnecessarily fall into traps that some of the customers or non-customers will set up. They will only be interested in pestering you and not really interested in buying any of your products and services. You have to learn to turn a blind eye to such people and continue with interacting with all your best customers.

You have to organize events for your best customers in order to understand them better and also keep them happy. The event needs to be good in order to keep the audience interested and educate as many new people about it as possible.

You have to focus on the offers that you organize for these people as that has the capacity to pull your audience in and keep them interested. Try to be as unique with your offers as possible and be creative.

There are other types of social networking sites that you can work with to promote your company. These include linked in, Google+

etc. All of these will help you reach out to a bigger and diverse audience. After all, isn't that what you want? A big audience that will keep your business going for a long time, so you have to put in an effort to be present on all the platforms that exist on the Internet and diversify your presence.

Conclusion

I thank you once again for choosing this book and hope you had a good time reading it.

The main aim of this book was to educate you on the importance of using social media to promote your company and products.

You will see that it is possible for you to increase your reach easily and get more and more people to like you and appreciate your efforts. The key is to do what works best for you and your company. As you read in this book, there are many strategies to pick from and you have to choose the best. Once you do, you have to implement it and promote your products and services. We looked at the 4 main social media platforms that you must master if you wish to turn your brand into a global image.

You have to keep with the times and adapt to all the new and upcoming technologies. You have to try and beat your competition and surpass them to reach your ultimate goal.

Through knowledge and experience you will see that it is progressively easier for you to advertise and market your company and make the most of your social media presence.

The next step is for you to implement everything that you read in this book and increase your company's reach.

Good luck!

YouTube Marketing

Table of Contents

Introduction:

Why Use YouTube for Marketing?

In 2017, digital marketing has taken on completely new forms. New methods of digital marketing are expanding as the industry becomes increasingly diversified across different platforms. This is both positive and negative news for online marketers. Depending on how well they can adapt and react to these changes, this can either be a hurdle or a great new opportunity for growth. It's going to be important, now, to utilize more than one media channel to both grow your brand and spread your content.

The Changing Face of Marketing:

Marketing was once easy, simple to attract viewers and create your own website, but now, due to the saturation of the market, the Internet is very competitive out there for those who wish to earn online. Drawing in an audience is no longer as easy as before since there are so many other sites competing to get audience attention. The following chapters will discuss some of the many ways you can stay relevant with your brand, or even create one from scratch, using YouTube. The final chapter will explore what to do beyond basic YouTube marketing. But first...

Why Use YouTube for Marketing?

Creating ads and content that attracts people means something different than it did before. Interactive content and virtual reality are of the risk now, and according to experts, YouTube views are going to become more important than ever.

- **To Create an Authority-based Image:** As mentioned, marketing is changing now, and YouTube has a lot to do with that. It's an essential tool for marketing no matter which company you're involved with or plan to become involved with. Uploading and importing videos is easier than ever, making it nearly effortless to add to your existing business plan. For companies, marketing using online videos offers a new opportunity that wasn't there before. Used the right way, and video marketing can help establish your company as being an authoritative voice on a matter.

- **A Huge Platform:** Another reason to take advantage of YouTube marketing is for the simple reason that it's the second largest online search engine that exists, offering your company a large platform for gaining audience members. The average customer is no longer satisfied with static marketing or physical ads and instead want engaging, aesthetic interactive content. Uploading video material gives users your ideas and business views, providing the aura of authority and expertise on the topic.

  When you share video content with strategies and tips, you can attract attention to your customers while providing them with value, building up your image as a respectable source. This is what keeps clients coming back time and time again, helping your brand grow.

- **Popularity:** YouTube didn't take long at all to become established as a hugely popular website, and this doesn't appear to be changing. They get over a billion unique audience members each and every month and endless hours of video content is watched each minute by these people. But most people just use the site to watch video content. They watch what they want to watch, maybe leave a comment and subscribe, and then watch something else.

Unfortunately, a lot of people who view the video content of others are not piloting the show, but are acting as mere passengers. Deciding to make your own channel on YouTube puts you into the driver's seat, making your content widely available for the millions of viewers on the site every day.

More Benefits to YouTube Marketing:

Still not convinced? Let's go over some more of the benefits to using YouTube for marketing.

1. SEO Advantages:

When you make a channel on YouTube, your search engine results are moving up already. You can then choose to use your YouTube channel and videos to bring more users to your blog, which also helps with your search engine results. In addition, videos do better than blog posts when it comes to search engines. Google offers a distinct benefit for YouTube on its search results because Google owns this website. A huge amount of the revenue earned by Google is actually on YouTube.

2. More Chances to be Discovered.

One very interesting piece of information about social media and shopping pages is that these websites are actually also search engines. YouTube is considered a search engine too, just as Google is. The search engine on Google lets you discover anything you want, but the search engine on YouTube will only seek out video content that is already on the site. YouTube isn't as big as Google is, of course, but it still gets a huge amount of unique visitors each month, as mentioned before. These users will be coming back over and over again and usually stay on YouTube for more than 10 minutes each session. This means that those who like what you make will come back to watch more of your content.

3. Your Content can Earn you More.

It's no secret that YouTube is great for sharing your message and promoting your blog, but did you know that YouTube itself is also profitable? People can actually earn millions from using Google AdSense with their existing video content. In fact, videos have made more than $100k, even those that are short (under two minutes in length). Envision it taking only a couple minutes to make a video, then earning that much from it! This is possible on YouTube, and it's an opportunity you shouldn't ignore.

True, the odds that you will hit it that big aren't very good, but you can still increase your subscribers and get up to thousands of views a day as time goes on. You can then expect to earn hundreds with YouTube. In addition, if your channel becomes very popular, you can start earning thousands each day.

There are plenty of books on this subject on the market, thanks again for choosing this one! Every effort was made to ensure it is full of as much useful information as possible. Please enjoy!

Chapter 1:

Setting up a YouTube Brand

Nothing brings more satisfaction than getting to a huge milestone when it comes to your YouTube channel and brand. The first milestone you can celebrate is the initial 1,000 subscribers. Making it to this many subscribers seems impossible to those who are new to the idea, but it's easier than you think!

Tips for Getting 1,000 Subscribers:

Throughout this book, we are going to give you tips on getting more and more subscribers until eventually, you can make it to 100,000 and beyond. As we will mention throughout the book, YouTube works best in conjunction with other websites and social networks and can be used as a catalyst to improve your business and audience. Let's cover some basic rules for enjoying a successful journey with YouTube.

Be Consistent:

To be successful with YouTube, you must post content that is valuable and do it on a consistent basis. This is actually quite challenging. Just looking around the website will lead you to think that every topic has already been made and even though that isn't true, you can take a topic that's already popular and put your own unique spin on it. For instance, a lot of techs "unboxing" videos exist, but you could make a "re-boxing" channel where you discuss different tech components as you place them in their box. This is just an idea, but feel free to be creative and come up with something on your own.

Give your Videos Smart Names:

Another challenging aspect of YouTube content creation is coming up with a good title for your videos. You may have an idea already, but it could be too long or complicated. Stick with the rule of keeping it as basic as you can, and YouTube can auto-complete the name. It doesn't matter if your video has a similar name to an existing video, but it does have to apply and make sense. If you are having a hard time coming up with quality names, there is a lot of extra information on the web about naming your videos successfully. Don't forget to do your research!

Make the Design of your Channel Appealing:

A lot of visitors are gathering a first impression of your brand and idea based on the design of your channel. This is how audience members can interact with you and learn what your channel stands for. Make an appealing header to showcase your company or brand. If you aren't very skilled with graphic design, font, or color selection, there are professionals online who are willing to help for a reasonable fee. In addition, you could ask a tech-savvy friend in your life to help you create an engaging channel style.

Try out Different Shooting Locations:

You can make your channel instantly more interesting by shooting videos in different areas. This could be challenging if you only have access to one area, but even changing up the background or going to a local coffee shop to record every once in a while can help keep your videos interesting to your audience members. This will keep them guessing and interested to return to see what you will do next.

Be Lighthearted:

It's no fun to watch someone that's too serious all the time, so try to poke fun at yourself and laugh on your channel every once in a while. Making mistakes openly allows you to appear more human to your audience, showing a different side than most of the other basic talking head content videos out there. Try to make it fresh, exciting, and spontaneous, so your viewers are surprised when they come to your channel. But in addition to this, always be yourself.

If you are the humorous type, make sure you show that in your videos. If you are serious, be yourself. What people will respond to and appreciate the most is someone genuine and relatable. You may not be liked by everyone, but you will be appreciated by the ones who matter.

Utilize the Custom Thumbnail Option:

You can increase your views and subscriptions on YouTube with this simple change. Just use a font that is popular on different photos online to get attention, then you can adjust the photo in Photoshop. You can make the title of the picture thumbnail something other than the video title so viewers can instantly tell what they are about to see.

Make Use of Annotations:

Have you ever been watching a video and see those little pop ups in the middle? Using these can be really helpful for making your content more effective. For each video, add an annotation to make subscribing to your channel easier for viewers. Sure, this does take a little extra effort, but as soon as you watch your rate of subscriptions increase, you'll want annotations in all of your videos.

Don't be Afraid to Ask:

This is a simple but often overlooked tip. You cannot get what you don't ask for! When your videos end, make sure you ask them to both subscribe and "like" your video and channel. This is a good time to also tell what audience members can expect from your channel. For instance, if you focus on home improvement project tips, you can tell users to subscribe to get more tips on the subject.

Be True to Yourself:

We already discussed this a bit above, but it's worth going into again. Always be yourself on your videos and channel! You can only be yourself and nothing else, even when you try. Some find being in front of the camera hard initially because you want to compare yourself to those who are skilled at being in front of the camera. Get over this by realizing that fakeness doesn't get you anywhere. Instead, allow your true personality to come through, and you will draw people towards you who resonate with who you are.

Make a Quality Trailer for Your Channel:

YouTube now gives video makers the chance to design their channel, including a trailer to cover what the channel is about. Most people only go see movies when they have already seen the trailer and can know what to expect. Your trailer should be no longer than a minute and a half, providing some scenes from the content you have on your channel along with reasons to subscribe. People will keep coming back if you put real effort into what you make.

The Third Most Popular Site Online:

YouTube is definitely number one in terms of video creation social networking, but it's also the third most visited site online. Yet not many people are taking full advantage of the social networking abilities it brings. Most users on YouTube don't even use the platform to create a brand, build their own presence, or make money. A lot of users haven't even put one video onto YouTube. If you're completely new to this idea, it's time to think about YouTube in a more serious way. Here are some reasons why:

- **Cross Promotion:** It's possible to cross promote using YouTube, using it to get viewers to your other websites or blogs. When your videos end, make sure you tell audience members that they can find you on your blog. Don't forget to remind them to subscribe, as well. YouTube is okay on its own, but where it's truly useful is as a supplement to existing blogs or websites. For this to work, you must engage in cross promotion on a regular basis for your channel.

- **A Full-Time Income can Happen:** Yes, it's possible, some have made full-time incomes from YouTube alone, especially through the AdSense program Google offers. When you first start out, however, make sure you are working on multiple income streams. AdSense really doesn't offer much money in comparison to the other strategies we are going to cover, but you still shouldn't ignore it.

- **A Chance to Grow your Business:** If you have videos on YouTube, people are likelier to watch your content. The more content you have, the longer they will watch it, and the longer they watch it, the higher the odds that they will end up subscribing to your content and finding your blog or other websites. Remember, the more, the better, but quality also matters!

The Importance of Optimized Titles:

Standing out from the crowd is the only way to have success on YouTube and one good way to accomplish this is by naming your videos something unique and offbeat. This will help people come to your channel due to curiosity. Having unique, quirky titles definitely plays a role in building up your views. However, in order to get as many views as you can, you have to know some SEO basics in terms of YouTube marketing. Here are some tips for that:

- **Title Keywords:** This was once a lot more effective for SEO than it is now, but it does still have an impact on your content. Crawlers on Google can't watch your videos the way they can read through text posts, so if you place your keywords into your video title, the bots on Google will be aware of the topic.

- **Utilize AdWords on Google:** In order to find out what search items are popular right now, utilize Google AdWords. Attempt to go for videos that have both low competition and high volume searches.

- **Have a Descriptive Title:** Although offbeat titles are a positive thing, you should also be giving your audience an idea of what your content is going to cover. Be unique but make sure that what your title says is relevant and makes sense in terms of what your video is about. This may take some practice to get good at, but you will if you keep at it.

- **Go for Shorter Titles for your Videos:** Keep in mind that Google shortens your video title down and puts "YouTube" before it, which takes up 10 more characters. Try to make your titles either 50 characters or shorter to maximize this.

- **Omit "Video" from the Video name:** There is no reason to include the word "video" when you are naming your video. This just eats up character space and doesn't help you in your YouTube search ratings. However, doing this might have a small impact on typical search results.

- **Customize your Channel:** When you are attempting to build trust with audience members using YouTube, you must use the options for customization on YouTube. Make it unique and fresh.

- **Be Professional:** Even though you are likely shooting your videos from your room, a professional appearance does matter in terms of gaining trust and respect from audience members. If your blog already has a following, try to use the same elements of branding on YouTube so others can recognize you across different websites and platforms.

- **Make a Bio:** Make sure you use the channel bio option YouTube offers to make your channel fully

- customized. Your bio information should be to the point and short. Add a link to your website, store page, or blog in video descriptions.

- **Placing Subscribe Buttons on your Website:** As mentioned, you should definitely include a link to your blog or website on your YouTube videos, but don't forget to add subscribe buttons on your blog to your YouTube channel, as well.

- **Keep your Videos less than Five Minutes Long:** YouTube has no shortage of in-depth, detailed videos and content, but the videos that convert the most subscribers are shorter than five minutes long. In 2014, it was found that most videos on YouTube are about 4.4 minutes long, so for those of you just starting out, this is an ideal number to shoot for. This will help you build up a loyal following.

As soon as you already have a following on your channel, you can play around with different lengths to see how people react, but at first, keep it entertaining, informative, simple, and short. It might be hard, but it's a worthy challenge and increases your likelihood of success with YouTube.

Having quality, custom-made art for your channel will go a long way in helping you get established with your brand online. YouTube has a service for channel art if you aren't skilled in graphic design.

Make sure you use a custom-made background head image with similar elements of design from your other website or blog.

Using the Best Tools with YouTube:
Plenty of useful tools exists to use with YouTube, including video creation tools, video promotion tools, and more. Making use of these will help you grow your subscribers in an organic way, and organic views mean more subscribers, over time. Here are some video marketing tools you can use to get the results you're looking for.

- **Bulk Suggest Tool:** This free tool was created by the Internet Marketing Ninjas and is an easy way to find out relevant keyword research data for your content. It will look through YouTube and Google's "auto complete" databases, expanding the terms you entered based on popular search terms. This allows you to compare search terms in Google fast with YouTube terms, better understanding the searching goals of your target subscribers and audience members.

  WordPress Keyword Plugin: In addition, there's a plugin for WordPress that you can use to look at keyword ideas using Google auto complete search terms. This is in your dashboard under the post editing options. Another benefit to this plugin is that it can give you ideas for what to create articles and videos about based on what is popular.

Do Research on Competitors:

Finding out more about your competitors will help you with your campaign for content marketing. Although this shouldn't get in the way of your creative ideas, analyzing your competitors will help you find out what works, along with tactics to help your viewers feel more interested and engaged in your channel. The best tool for this is the YouTube Analysis feature on BirdSong Analytics. This tool allows you to pay as you go. It only requires a channel name to use, and this handy tool will give you lots of useful reports and stats, such as:

- Your competitor's channel's best upload time (in terms of getting likes).

- The ideal day to upload for getting likes, according to your competitor.

- When you should upload to receive audience comments, according to the competition.

- The best day to upload videos in order to get comments, according to your competition's channel.

- The way the length of your videos can impact viewing figures.

- The way the length of the videos can impact viewer engagement.

- Commonly used terms in video captions, and more.

It also allows you to download a spreadsheet from Excel that lists all of the videos from your competitors, including upload date, description, title, duration, and day of upload, along with how many views, comments, and likes each video has. Mess around with conditional formatting and filters within the spreadsheet, and you can learn even more about competitor tactics. Make sure that you look into more than one channel before attempting to use these reports for your video creation.

How Can You Get People to Share Your Content?

Each time you put a video up on your channel, you have made an asset toward your content that must be promoted properly across your channels of social media. Making use of the RSS feeds on your channel, you may share your content to many different channels of social media using a channel that is semi-automated. Here is a tool that can help you with that:

DrumUp:

This handy tool allows you to schedule items on your feed with just a simple click, putting them up on your social media. This tool allows you to add more than one Twitter account, more than one Facebook page, and also your LinkedIn profile. You are able to promote your video content in all places in a much more efficient manner. It also allows you to:

- Schedule your videos fast using their one-click schedule button. This distributes them across your different social media platforms.

- Preview what your snippets will appear like on social media once they have been posted to the platforms.

- Authenticate your various accounts on social media.

As you can see, there are plenty of tools to make use on for improving your YouTube channel, stats, and views.

Don't be afraid to do some extra research outside of this book to find even more sources.

Chapter 2:

How to Increase your YouTube Views and Earnings

I f YouTube is just a place you go to listen to music or watch cute cat videos, you are seriously holding yourself back on the advantages it can bring you. This can be used as a powerful tool for business to strengthen your online presence and make your brand better-known.

The Three YouTube Channel Types:

Countless resources exist on the topic of YouTube and the different ways you can use it for maximizing your business (and income), but this section will focus specifically on the advantages of branding your channel.

1. **Custom Branded:** This type of channel allows for a website-like interface. However, this probably won't happen for you unless you have a gigantic budget.

2. **Branded:** This option offers the ability to make an experience for audience members that are more

customized than the user channel, which is the most basic.

3. **User:** This is the simplest type of channel. Your videos come with analytics, and you can only customize background images and colors.

Benefits to Setting up a Brand for your YouTube Channel:

Setting up a brand for your YouTube channel shouldn't be done just because. However, you can do this (and should) if you have clear objectives, goals, and a strategy to implement. You also need to already have some videos, so if you aren't to that stage yet, come back to this section once you are. If you do this right, you can enjoy the following benefits:

-   **Consistency for your Customers:** When your channel on YouTube is branded, your page can be customized, making your page more consistent in terms of how it compares to your website. You will be able to add background pictures and banners, along with additional tabs and customized sections for content. In addition, a YouTube channel that is branded will give you more optimization opportunities, helping your business get more visibility.

    Remember that YouTube is number two in terms of the world's biggest search engines. YouTube's tagging option lets you add keywords to every video, communicating the subject of your content to search engines. In addition, YouTube comes with its own version of Analytics for you to monitor the performance of other successful videos, visits, and more.

Expands your Outreach Socially: One of the best benefits of YouTube for marketing is how well it integrates with other platforms, including Tumblr, Pinterest, and Twitter. You can add links to those

- channels on your brand banner, letting audience members find your other content.

Need More Views on YouTube?

YouTube is a great method for getting your expertise and message out to a wider audience. But no body is going to find out about your expertise and message if no one looks at your videos. As the views accumulate, more and more people will learn about you, and you will then get more sales, subscribers, and followers across other platforms. As mentioned, you need to be consistent with your video publishing schedule. When people know to expect a video from you each week or month, they will come back to your channel to see it. Here are some other tips for getting more YouTube views:

Have Plenty of Content: Your very first YouTube video might be quality enough for a single subscriber, but your audience isn't going to just watch the same

- video every time they go online You must make new videos on a frequent basis, so your viewers and subscribers have content to choose from.

- **Hint at other Videos in your Content:** You should feature other video content of yours within the

last seconds of your videos. After YouTube viewers watch videos, some may lose their attention span in the process of searching for something else to watch. When videos end, more options come up. As soon as an audience member sees these other options, they might go to that, leaving your channel.

Plan for this by including some links to similar videos on your channel at the end of each video you make. This allows your viewers to find more of your content. When other options are simpler to access, you will get more views.

- **Let Viewers Know you:** When audience members find your videos and see them initially, they aren't going to know anything about your videos, channel, or brand. You can fix this by telling them your name, what you do, and more about your business or project at the start of every video.

Utilize YouTube Comments: When you comment on people's YouTube content, people will see your profile name and create YouTube channel backlinks.

- These links will lead to a higher number of views for you, eventually.

Make a FAQ Video: Every successful YouTube video creator should have a FAQ video that allows you to answer frequently asked questions your audience might have. Ultimately, your audience members are the ones who choose to link to your content on their Twitter, or who they will share your videos with on Facebook.

- When you make a video that involves frequently asked questions and their answers, it increases the likelihood that people will share it.

Making the Most of Video Descriptions:

Returning to the SEO involved in content on YouTube, the descriptions in your videos shouldn't ever be neglected. This section will allow you to be more easily found online and will also let people know what you create videos about. However, you don't have to overdo this part. A super detailed video description doesn't make a lot of sense since it won't all show when someone loads your video. Just as we mentioned with your video titles, make sure you include keywords for the video in the video description, but keep it simple, authentic, and natural.

Using Meta Tags:

You can utilize Google Keyword Planner to get some ideas on which keywords you should include in your content. These keywords can then be added to the videos you make which will allow you to be more easily found on YouTube and Google. Keep in mind that trying to use too many keywords will have a negative impact on your views, but that a few well-placed and researched keywords can aid you in your ratings and rankings on YouTube. A low number of videos doesn't always mean the content is poor, but it does mean that discovering it is harder and less likely.

Metadata plays a large role in helping your content show up when people search. Look up some videos that are well-converting to see which tags they are putting on their videos. This should give you a few ideas but make sure you don't copy them as this will not help you at all. Remember to be unique and creative!

End in a Positive and Memorable Fashion:

No matter what type of videos you make, they should always be ended in a positive and memorable way.

- **What to Include in the End:** Just as you should always have a standard intro to your videos, you should also make a recognizable outro. Ask the viewers if they liked the video, ask them to subscribe and leave comments and tell them to look at your blog or website.

- **Don't be Afraid to Ask:** No matter which approach you take, keep in mind that those who don't ask never get told yes. Make sure that the ending to your videos is always confident and that you humbly thank your viewers for following you. End smiling and make it so that your audience wants to return to see more of you.

Collaboration with Other YouTube Content Creators:

Lately, a lot of YouTube video creators collaborate with each other, but why is that? Collaboration is beneficial for everyone involved, including the audience, you, and the person you work with.

- **Reaching Out:** Creativity is a constructive process, and when you see other YouTube video creators as competition and nothing else, this will only hold you back. Why not try to reach out to them and work together instead? That way, their success is your success and vice versa. This will help you access a wider audience, and your collaborator will enjoy the same privilege, then your viewers can all get some extra value.

- **Keeping it Fresh:** Working with other YouTube video creators is also a way that you can keep your content fresh and interesting. Surprise your viewers by offering them something completely new!

Always Interact with the Viewers:

Social media is all about interacting and connecting with other people and how much you can show them you care. When your viewers can tell that you actually care about who they are, they will return the favor for you. Listen to what they request, interact with them, and read and respond to comments on your videos. It's true that YouTube has plenty of anger and backlash in the comments sections, but ignore the haters and thank your faithful viewers by giving them the time of day. This will lead to respect and trust.

Chapter 3: How to Increase Viewing Time on Videos

Did you know that people use YouTube for learning about services and products more than other social networks, such as Facebook? What this means is that for those of you who review or sell products, YouTube is a must for your business marketing. Most customers wish to both hear someone explain and see what a product looks like, so they know what to expect. YouTube can provide the answers to these specific questions. If you still are trying to decide how to use YouTube, you can review products through affiliate marketing.

Promoting your Own Items:

Another method for generating income using YouTube is the promotion of your own items or products. A lot of costly training programs have come out online. Certain ones cost $90, and others cost $900. You can run a promotion for your affiliate link for a costly product or training course using your own YouTube videos, and this can earn you big bucks. If you were able to sell a training course for $900 and earn $450 for every sale you make, this would add up a lot over time.

It's true that AdSense can help you make money using YouTube, but that's not the only way to do it. Actually, for a new YouTube video creator, promoting items and products is a faster method for earning a good income, so I recommend starting with this if you're new to the game. True, ads can make some money each day, but affiliate links are better at first. Remember that you shouldn't only stick to one method. Promoting your items, using affiliate links, and AdSense should all be explored.

YouTube Allows Anyone to Begin:

YouTube is great because even amateurs can have their own channel. Out of every method for creating videos online out there, YouTube is by far the simplest.

- **No Learning Curve:** No technical skill is required to do this. You shouldn't have to worry about learning a bunch of technical stuff when you want to make videos with value in them.

  Handheld YouTube Video Creation: YouTube allows you the easy way of sharing valuable content and makes it as easy as just uploading your file and describing it, then hitting publish. This can even be

- done from a handheld device. Use your iPad to upload content and stop worrying about whether you are doing something right. Just focus on creating great videos.

Increasing Viewing Time on YouTube:

A lot of people are still stuck on finding methods for getting more YouTube clicks on their videos, believing that this is what makes content go viral. While this was once true, it no longer is. Now, the length of time a person watches the video counts more. What this means is that if you got a million views, but for each view, people only stayed at your video for seven seconds, then it's not considered engaging enough for YouTube to note or promote your content. Now, how long you can keep someone watching is what counts as viral on the website. How can you keep people watching your videos? Follow the tips below:

- **Aim for More Subscribers:** The people who are subscribed to your channel will be the main people staying for your entire videos.

- **Make your Introduction Professional:** The way your introduction is will control how people react to your videos, as a whole. If your introduction is professional and friendly, people are going to like your content more; it's as simple as that.

Shorten Your Content: How do you think people would rather get their information, by sitting through

• something for an hour, or for two minutes? We've said this before, but the shorter your videos are, the better.

Get More Than One Camera Angle: Using more than one camera angle helps a lot. If you do this,

•

messing up is not a problem because you can just keep your video going with the alternate camera. Right when you realize that you were rambling on a bit too much during your video, you can cease the first angle recording and then just switch to recording from the other one. Keep in mind that when people ramble in YouTube videos, people usually don't keep watching them.

The Content Must be good: Keep in mind that your content has to be good and that no amount of special effects or expensive equipment can make up for the lack of valuable content. People won't watch a video that is bad just because the factors above are

• included. You have to always work on making your videos better and better. Sure, they might be good, but how can they be great?

Prove that People like you: The next step involves showing people that you are liked and popular. Most people think that viewership really affects the way

- YouTube puts videos into rank. What this means is that most believe popularity is counted by views on your videos. If one of your videos has 3,000 views, people are more likely to watch it all the way through since others have already watched them. So even though keeping people on your video matters more than the clicks you get, clicks do matter too.

Lengthening how long people stay on your videos will put you at a higher rank on the search engine of YouTube, where you can access more audience members who could potentially subscribe. In other words, don't skip this part if you want a successful YouTube channel!

Video Challenges and Giveaways:

Everyone likes giveaways. You should offer audience members a reward for being dedicated and engaged with your channel. A video contest or free giveaway is a way to reward the ones that are already following you while drawing in potential subscribers and followers. Giveaways can include some of the following items:

- T-shirts.

- Tech gadgets.

- Hosting subscriptions.

- CDs.

- A free e-book.

These are just a few ideas. Whatever you decide to give away, your viewers will appreciate the opportunity to receive something free and will be likelier to share about your channel to their friends. This is free, viral promotion for your YouTube channel. In order to organize a large giveaway, a lot of hosts on YouTube call for each audience member to follow their profiles on social media in order to be accepted for the drawing. This approach is recommended. It's a good idea if the item you are giving has something to do with your channel topic, but you can do it even if it isn't.

More on Cross Platform Marketing:

Promoting across different platforms is not just recommended, it's a necessity for YouTube marketing. Now that we are well into the social media age being active and present on more than one social profile is a must for existing. If you want your brand to get recognition, you have to be discoverable, meaning activity on social media is required. Make a profile on Google Plus, one on Twitter, and of course, one on Facebook if you aren't already signed up.

In addition, don't forget about the lesser-used platforms such as Snap Chat, Instagram, and Pinterest. If you don't like the idea of having to dedicate a lot of time to this pursuit, try looking into auto-scheduling for your posts which will keep you organized. You can also try out Google Ads and Facebook Ads to further your marketing. In the modern age, this is how you build a successful business and brand. Visibility across the web sets you up with a ubiquitous presence.

Aggressive Marketing:

In order to get more subscribers and make an income with YouTube, you have to be aggressive in your approach to getting successful. The idea that building it up first and that the followers will flow in naturally doesn't always work with social media due to the amount of competition present. You have to actively promote yourself and your channel
as much as you can. A YouTube channel that is growing in popularity will help you stay motivated to keep going, making valuable video content for your viewers.

Tell your friends about what you're doing, asking them to subscribe if they want to. Keep in mind that pestering people will have the opposite effect that you are going for. Get online and make personal connections, then ask them to subscribe and like without being annoying about it.

Chapter 4:

What Matters Most for Getting Subscribers?

W hen you are building a business, you have to draw in your audience using more than basic sales tactics. Trust matters a lot when you want to draw in customers. If your customers don't respect or trust you, you aren't going to have any luck. Therefore, you have to take the time to build up trust and rapport with the audience and customers. When you do this, you will have an advantage over other businesses and help your list of subscribers grow. When you upload video content to YouTube, you are giving your customers reasons to evaluate and look at your products and services.

Credibility and YouTube Marketing:

Credibility arises when someone has all the information needed. There isn't a more direct, more easily absorbed, or simpler way to show information than using YouTube videos. Just 20 seconds of content is a lot more engaging and offers more value than a whole page of text or marketing content.

- **Showing Satisfied Customers:** The best way to build up your credibility and trust using videos is to put videos up on your channel of people recommending your product and proof of client engagement. For potential customers, a satisfied client giving their opinion is the most convincing piece of evidence that your brand is worthwhile.

- **Winning New Clients:** Proving the way you respect and treat your existing customers is going to help you gain new ones. Videos which show customers enjoying your brand or showing how good it is will engage new viewers and subscribers. Testimonials on your channel will build up trust by demonstrating and displaying the value in a way that is engaging and meaningful in a way that other ads could never be. Small or large, any business needs to use video content to build and establish their business and products.

One simple method for doing this is putting a video up that has instructions or information steps, targeting your audience with success and accomplishing a set of actions or tasks. As newer social networks, such as Blab and Periscope pop up, older websites seem to pale in comparison. YouTube is an example of this, in some cases. Although it's still wildly popular, people might fear that YouTube isn't going to be as popular pretty soon.

The fact is that YouTube has existed for very long, leading people to believe that newer websites make more sense to put their time and effort into. Although the newer sites are valuable and useful in their own way, it isn't smart to exclude YouTube from your current strategy of marketing.

The Three Most Important Factors on YouTube:

If your desire is to create a YouTube channel that is successful and shows your niche, you must already know how to create a successful brand and channel. To be more specific, you have to be aware of the three consideration that is most important on YouTube.

1. Your Quantity of Subscribers:

A lot of interested, quality subscribers and viewers is your ideal YouTube audience. Quality, interested subscribers, means anyone who is already interested in the niche you are involved with. They are already subscribed to other channels that are somewhat related to the niche you're in and like watching a lot of videos about that subject.

> **Appeal plus Marketing:** Attaining a higher number of subscribers means creating a channel that is appealing and combining that with the correct form of marketing. It does help to have an existing audience outside your videos, but you don't need this. It's possible to leverage the audiences of other people's challenges by interviewing experts in your niche, leaving comments on

- related video content, and more. However, when you choose to try to go leverage the audiences of other channels, you can only enjoy success if you choose to have a "win-win" attitude.

- **Commenting on Videos:** You can get more exposure by commenting, while your comments offer more social proof and exposure for the videos. This is a win-win situation for you and the video creator.

  Interviewing Experts: Seeking out information from existing experts in the field will help you get more exposure and credibility (especially if the person also does promotion for the interview). This offers a win-win for you and the expert. Other methods exist for getting a higher number of subscribers, but the best approach is to find a couple that work best for you, then focus on those.

2. How Long the Videos are Watched:

YouTube has intentionally emphasized that when they think about videos viewed, they are also considering how many minutes of video are watched. This is partially to help against spam since people can just purchase views. These fake views are dishonest and make video content appear better to possible audience members. But these videos will likely not appear high up in the search results of YouTube. To get a higher rank, you need to make videos that people want to watch all the way through, or at least, most of the way through. You can do this with two different methods:

- **Creating Compelling Content:** The first way to get people to watch your videos all the way through is to create compelling content that makes people want to watch. How is this done? You can check out what your successful competitors are doing and look at what works here. Next, check out the retention rate on your own videos to see what works. Don't check out the retention rate of your videos until you've gotten past 100 views. If your retention rate is based on just eight views, this isn't enough information to form any conclusions.

- **Making One Longer Video:** The other method for increasing the length of time your videos are viewed is to of course create a valuable video. However, how long your video is matters, as well. We have already mentioned that when it comes to YouTube subscribers, having shorter videos is better, but this is an exception. Retention rate determines how long someone sits through your video, but how many minutes they watch for matters more. If you want one specific video to get great ratings, you can make it a bit longer.

You can do plenty of planning to prepare for your video if you want to make one specific piece of content longer. For instance, you could do an instructional video on how to develop an exercise routine and make it 25 minutes long instead of four. Using advertising's help, you can get people to watch this video for thousands of minutes. Your average watch time for your videos will go up due to having a longer video that people watch for longer.

It won't matter whether the longer video and another shorter video get the same amount of clicks, but it will matter that people spend more time watching the longer one than the shorter one. Make sure you make compelling content, but for some of them, make longer videos to add to your overall minutes-watched score.

3. How Consistently you Upload Videos:

If you are more consistent with uploading videos, you will also be likelier to gain attention from those subscribed to your channel. A lot of people subscribe to a channel and then forget about it unless you make sure you are consistent with putting out content. The thing about publishing and uploading videos is some audience members will have email notifications turned on, so that each time you put something up, a group of viewers will know. Your subscribers are likelier to remember you if your videos come out on a regular basis. If your videos are uploaded once a week, on Friday afternoons, your audience will be more likely to remember that you and your channel exist.

Some of your subscribers are going to look at your channel regularly to find what you have released most recently. These audience members will usually engage with, watch, and even comment on your video. When they do this, your video will go up in the rankings on YouTube, bringing more traffic through the search engine, too.

The audience members are the ones who will get this momentum flowing for you. But for them to be able to do this, they have to be able to remember your channel, meaning consistency of uploading is key for you.

The Powerful Video Platform:

In our modern world which is increasingly full of social networks, YouTube remains a very powerful platform for your videos. Since we have covered the factors that matter most for building up your subscribers, you can build your channel and videos using this information. Obviously, the most important factor of all is that you create valuable content, but you also need to know how to do this as it relates to SEO on YouTube. Experimenting with plenty of ideas will allow you to find out what works best.

Chapter 5:

How to Win and Keep Subscribers on YouTube

Videos are leading the scene of content marketing as of this year. YouTube is in the lead as far as video blogging goes, along with video marketing and video sharing. The platform is free and offered by the giant search engine, Google. People absolutely love YouTube and use it all the time. Twitter and Facebook have also tried to join the game of video marketing recently but as of yet, have only a tiny fraction of the impact of YouTube. If you already have a channel on YouTube, you might be curious about how to draw more subscribers to your brand as well as increasing how far your videos reach.

Since YouTube has at least a billion visitors each month, a gigantic potential exists for building up your audience, each time you make a new video. This can be a fashion design review video or a skateboard trick tutorial. Either way, YouTube is where people go to find the videos they want. Plenty of brand new YouTube stars are being made every day, so for those seeking publicity and online marketing expansion, it's a must. YouTube is the best way to expand your reach across the online world as a blogger, increasing your connections with readers.

More Tips for Getting Subscribers:

The chances for bloggers to make use of YouTube and bring more traffic to their site is high. In order for this to occur, however, you first need to know how to bring more subscribers to your channel. This detailed section will show you just how to do that. If your main business goal right now is to build up your channel, these tips will help you see tremendous growth within a matter of months.

Make a Script and Plan:

The initial step to making your journey on YouTube successful is to have a plan for your channel and video content. The next step is then to structure how your videos will go. Choose what you like to create and work on getting the skills related to that. Although we mentioned drawing inspiration from your competitor's channels, don't copy what they do. Sticking to what you are passionate about is key for enjoying success with this platform.

> **Writing a Script:** Your video creation will go much smoother if you actually write out a script since this will keep you on track as you speak and make organization much easier. When you make a script to stick with, you will stay on task instead of getting lost in rambles. This will keep your events flowing perfectly and make your

video focused. When you write your script, include the
- main points to cover, calls to action for your audience (such as subscribing or liking the video), how many words you'll say, and what actions you'll perform while on camera.

- **Identify the Target:** You need to also identify the target audience, structuring your script on their current understanding and level of information. If you're doing tech videos, are you speaking to tech beginners? Are you talking about Americans or non-native speakers of English? Are they intermediate or experts in the field you're discussing? Are you trying to be informative or funny to suit their tastes? Using the right language for your target audience is a must!

Making Entertaining Videos:

We've already discussed the importance of making videos that are valuable to your audience. This is an obvious one, but your content has to be informative, engaging and also entertaining for viewers. And this has to be the case throughout the whole video, not just the intro or ending. If you lose your hook right in the middle of your video, you're going to lose viewers.

Informative Video Content: Content that is most valuable to viewers is both informative and entertaining instead of being just one or the other. With all types of

marketing, this is the standard, but with videos, it's especially important. Think about it, if someone wanted to find something out just for its informative properties,
- why wouldn't they just read an article or get a book on the subject? They are watching videos to receive information in a new, more engaging way.

- **Evergreen and Burst:** More specifically, try to upload a mix of evergreen and burst videos. The burst videos are popular for a short time to get you hits in the short term, but won't stay popular past a certain amount of time. Evergreen content, however, will stay relevant for a long time and get you valuable archived YouTube views. If you can, attempt to make content that is mostly evergreen. If you have a hard time being comfortable in front of a camera, do screen casts, which involve giving information with other images shown.

  However, whichever type of videos you focus on, always ensure they are both valuable and highly engaging before you publish them.

Aim for making More Videos:

This might be simpler to say than to do, but it's very valid for the future success of your channel. One of the main reasons people subscribe to YouTube channels is because they appreciate what they see and are expecting more of it.

Don't be Forgotten: Subscribers on YouTube won't usually appreciate channels that are dead or don't make a lot of videos. In our modern age, especially, consumers are always looking for more entertainment, and you must be ready to follow these demands to stay relevant. As we've already

- mentioned many times in the book, consistency is the only way to create a valuable, lasting relationship to your audience.

Stick to Your Schedule: Try to release your videos according to a strict schedule. One each week is ideal, but you should go for a minimum of one each month. Stay with that schedule and try not to upload anything between, which could hinder your YouTube reputation.

- People enjoy watching TV series because they can look forward to episodes on certain days. Following this structure will help you be more appealing to viewers.

Be Ruthless in your Editing:

With your YouTube videos, you need to be ruthless with your editing. The famous photographer, Thomas Hawk, says that for each picture he publishes, 10 have been rejected. This should apply to all areas of editing, videos included.

Don't Rush the Process: Be ruthless with your editing to make sure that what you publish is only your

absolute best. If you rush through this for the sake of
sticking to your publishing schedule, it's going to hurt
• your business over time. Instead, adequately prepare
and give yourself plenty of wiggle room as far as time
goes, so you can select the best content.

Ask a Friend: If you can't tell which shots of you are
best for your final video, ask a friend to watch with you
• and help you decide which ones are upload-ready.

Do Plenty of Recordings: Whenever you decide to
shoot a video, do plenty of recordings and only select
the best ones. Take a lot of shots until you find one
that you are confident about. You can edit with Adobe
Premier for Windows. The right editing tools will help
a lot with this and depending on how serious you want
• to get about it, you could hire someone to help you
edit.

Always Explore and Experiment:

This is meant to be a fun experiment, so don't forget to keep
exploring to find what works best for you, personally! Someone
else's methods may not work the best for you, so find your own.
Experiment with backgrounds, camera angles, thumbnails, and the
rest of the methods given to you throughout this book, then track
the changes you make and how they impact your audience's
reactions. Stick with your business and brand, always staying true

to it. Keep in mind that creating something valuable using YouTube requires perseverance, time, and effort. But if you're committed, you'll see the benefits in no time.

Chapter 6:

Creating Winning, Profitable YouTube Content

If you're one of the millions of users on YouTube every day, watching videos, it's about time you started creating your own content. Instead of being a passenger, you can be a pilot. YouTube allows people to create their own videos, upload them, and become discovered on a larger scale than was ever possible before. In fact, many celebrities have been discovered this way. The revenue and viewership are small at first, but if you stick to your plan and promote your channel, it could become one of the popular ones on YouTube, bringing you a great income.

Getting people's attention is not the same as keeping it, which is why YouTube has changed its system of ranking videos. The videos that hold the attention of viewers the longest are going to get higher rankings than the ones who get people to click on them. This should help you feel more motivated to create quality videos. So how can you improve this to make your channel more successful? Follow these tips for creating a stronger engagement with your viewers, improving your ranking.

Don't Game the System:

There's no point in trying to gain a higher ranking by bypassing YouTube's algorithms. As soon as someone believes they have beaten this, it could change. Instead, focus your efforts on researching keywords, crafting the right titles, and making videos consistently, as mentioned before. Producing valuable content consistently will give you the results you want.

Know Who's Watching:

Knowing your audience is the next key to success on YouTube. You should know how old the people watching are, what they enjoy doing, which words will get them to pay attention, what they want to learn, and which benefits they're hunting for. This is simple stuff, but it is often overlooked by people. You should never assume that you already know the needs of your audience without doing proper research to find out. Here is how to maximize your relationship with your viewers on YouTube:

- **Offer Incentives for Staying:** Your viewers should know within under 30 seconds why they must stay on your channel. Let them know what you can offer them and then prove it throughout the rest of your video.

  Pre-production Planning: Great videos don't just occur by chance. They need pre-production planning. This includes a shot list, a list of props for each scene, a story board, and any additional

- footage needed to bring your video to life in the editing stages.

- **Be Energetic:** Those who are enthusiastic on their videos are far likelier to keep the attention of their viewers than someone who is dull and monotonous. You can study this by watching stand-up comics, watching how their energy has an impact on viewers. Try to talk louder than you usually would and gesture a little more exaggeratedly. This might feel a little awkward at first but usually, results in a more exciting end product that will draw viewers in more.

Don't be Afraid to Take Risks:

As soon as you know who your audience members are, go with your intuition and don't be afraid to take calculated risks. Every niche has a bunch of untested assumptions, so test out common beliefs. This will show viewers that you aren't afraid to think outside of the box and you may learn a few things along the way. In addition, make sure you are paying attention to descriptions and headlines in your videos. Your content might be great, but if it doesn't have the correct keywords in the tags and headlines, people won't find it! Use the keyword tool on Google to find out whether the words you selected are popular or not. Try to find a minimum of 10k searches globally that are "low competition."

Chapter 7:

YouTube Marketing, the Next Step

T he far reach and convenience offered by the Internet has offered countless people the chance to earn their living digitally. This involves the chance to monetize nearly every opportunity, talent, or skill. And as with any other venture for making money, plenty of misinformation is out there on the topic, including YouTube marketing. More specifically, the name YouTube comes up a lot in articles that talk about earning income from home. But although you can make income using this platform, it usually doesn't happen in the way people believe it will. Actually, to be successful with this, you must establish a strategy that is more sustainable, which requires digging below the surface.

Earning Income with YouTube- the Challenges:

The myth of YouTube says that all you have to do to earn money is put up some videos, bring viewers to your channel, and then make money from the advertisements on the videos. Does this sound effortless or too good to be true? That's because it's both. This is the story that everyone repeats, but the truth is that ad revenue won't bring you the income you seek. Even for advertisers that do pay a lot for the promotion of their services using ads, only a tiny amount of this goes to the creators.

To put it another way, for each million views, you won't make more than a thousand dollars or so. And getting to this many views is a lot harder than it sounds! Thankfully, ads on YouTube are far from the only way to generate income for yourself, as long as you are willing to come up with a plan and work hard.

Other Options for Making an Income on YouTube:

To get started with making good money on YouTube, you shouldn't view it as a medium that is monetize-able on its own and should instead view it as a useful catalyst. The best way to earn an income using YouTube is to use its gigantic network of users. Follow these tips to do that:

- **Using Shopify:** As we've mentioned a few times, YouTube is the world's second biggest search engine, after Google. Considering this from the viewpoint of marketing, it just doesn't make a lot of sense to not take advantage of that. One great opportunity for making money is using YouTube as a platform for selling your own merchandise. Perhaps you already have some merchandise to try to sell. You can use Shopify to set up your storefront online, then make videos related to the niche your product is in. Once the videos are over, include a link to your shop's landing page.

This process involves more than just that, but a good product can be sold much easier with simple videos. You can product how-to videos or simple question and answer style content.

Using Yondo: If you have a goal to earn money from your YouTube videos, you can take advantage of a much better choice than relying on ad revenue. Start with making a channel on YouTube and gaining subscribers. The goal here is to build your brand up while engaging with your viewers. As soon as your reputation is solid, you can start directing traffic to your blog or selling page where you can sell premium video content. You can use solutions such as Yondo to do this, which will allow you to make a store that sells video content.

Your free content on YouTube will pull people in and those who want to see more will have the chance to buy more. Your videos can sell as rentals on a pay per view purpose, subscriptions by the month, and more. You will set your own price and don't have to split the profits with YouTube.

Affiliate Linking: Affiliate marketing is a highly popular method of making money online. The problem with that is that a lot of marketers doing this aren't going about it the right way. Rather than using your blog to get traffic to your affiliate links, create a channel on YouTube instead. This is much more engaging and

- entertaining for people. The internet appears to be moving toward a preference for videos over text, meaning that you won't get left in the dust.

Using YouTube for Sponsorships: The YouTube video creators who have enjoyed the most success

- include advertisements and sponsorships in their videos. The makers of these videos have found these opportunities on their own, more often than not. The best thing about this is that YouTube won't take a cut of your earnings. In addition, you can make your own contracts based on your goals and audience size instead of working under YouTube's terms.

Usually, how much revenue is generated through this method depends on how hard you work and almost always ends up being more than ad revenue from YouTube alone. Keep in mind that you can still work on getting ad revenue in the meantime, drawing on two different sources of money from just one video.

Consider Live Speaking: You can leverage your reputation on YouTube by transitioning into live speaking. If your channel is involved with a specific audience or niche, conduct some research regarding related industry events or annual conferences that need speakers. You can then pull together your best clips and YouTube statistics, creating a pitch and package to show

- to event directors. Speaking engagements, such as these, can earn you a lot of money, even with just an hour-long talk. Ensure that you aren't missing out on this opportunity, which can then grow your audience even more.

The tips above help you to use YouTube as a tool, not as the "end all be all." YouTube can be one stepping stone on the way to success for you and a catalyst to make your dreams come true.

Think Creatively:

As with any other venture, you should never be afraid to think creatively. The guidelines in this book are just that; guidelines. They aren't rules that you must adhere to if they aren't working for you. Will you be successful with earning money on YouTube? That's up to you.

Conclusion

YouTube is becoming one of the most powerful communication platforms and is far more engaging than other ads. People like to watch videos. It's easy and often entertaining for the audience members. Videos allow you to connect more personally with your followers, offering a hands-on approach to marketing and your brand.

For your marketing to be successful, engagement cannot be skipped out on. Websites that allow you to show videos give you so many opportunities for engagement. The viewers can lead comments on the videos, interacting both with you and each other. They can see your face and hear your voice, making you more trustworthy to them.

Don't forget to craft a memorable intro and outro for your videos. This not only helps you with your brand but helps your video seem more appealing and entertaining to your audience members. This will serve your business and brand, offering the video some professionalism, just like a TV show's main theme song. In addition, an appealing video intro keeps your viewers wanting more. Your outro is your chance to encourage viewers to subscribe!

YouTube humanizes companies, making them more helpful, professional, and approachable. Adding this to your business strategy will help you no matter which industry you're in. If you're ready to take advantage of this, follow the guidelines given to you in this book and if you stick with it, you'll enjoy success with YouTube Marketing.

Finally, if you enjoyed this useful guide, please take the time to leave it a positive review on Amazon. Thank you again and good luck on your YouTube marketing journey!

Blogging

Table of Contents

Introduction

I want to thank you and congratulate you for purchasing this book...

This book will teach you how to earn money through your blog.

It will help you become a skilled blogger. It will provide you with tips, tricks, techniques, and strategies that you can use as a blogger. It is a comprehensive guide: it will teach you how to start a blog, write content, promote your site, sell ads, join ad networks, and many more.

It contains screenshots, detailed explanations, and step-by-step instructions. Thus, after reading this book, you will be able to create a blog that generates income. It will also help you enhance your writing and marketing skills. If you want to be a skilled blogger, this is the book that you need to read.

Thanks again for purchasing this book, I hope you enjoy it!

Chapter 1:

How to Establish Your Own Blog

This chapter will teach you how to create your own blog. It will provide you with screenshots and detailed instructions. Thus, you won't experience any problems setting up your blog even if you're not familiar with the process.

1. Search for a Domain – First, you need to find the perfect domain name for your blog. As a general rule, your chosen domain name must be directly linked to the topic/s you will write about. For example, if you're going to write articles about the game of chess, you might consider "chesslessons.com," "letsplaychess.net," or "thechessblog.org."

Since websites can't share a domain name, you need to make sure that your chosen domain is available. Visit www.domize.com and enter the domain name you want to use. The website will help you check the availability of the domain names you selected. It will also give you numerous suggestions, which can be handy if the original domain name you entered is already taken.

2. Purchase the domain name and get a hosting package for your website – Countless hosting service providers are out in the market. Some of the best webhosts are GoDaddy, Network Solutions and Hostgator. To keep this book simple, let's assume that you chose to get the services of Hostgator.

3. Install the WordPress software – After purchasing the domain and hosting package, you must install the WordPress software on your new site. This process can be technical so you are advised to contact your webhost. Basically, you need to contact Hostgator. Once a customer service representative answers your call, inform him/her that you need to install the WordPress program. Then, follow the representative's instructions. Get the login credentials for your WordPress account once the installation process is complete.

4. Access the WordPress system – Launch your favorite web browser and type the following address: www.domainname.com/wp-admin. Just replace the "domainname.com" with your own domain name. Then, enter your login information (i.e. username and password). Once you're logged in, you may start publishing posts and/or customizing your blog.

The Basics of Blogging

This part of the book will explain the basic concepts related to blogging. Read this material carefully since it will help you become an effective blogger.

Blogs

The term "blog" originated from the phrase "web log." In general, blogs are websites that contain viewable materials (e.g. articles). Most "bloggers" (i.e. the people who have blogs) use their blogs as online diaries.

Several years ago, bloggers needed to be familiar with one or more programming languages. Fortunately, blogging platforms (e.g. WordPress) and software producers turned blogging into a simple activity. These days, you can create great blogs even if you don't know any programming language.

Some Thoughts About Starting a Blog

Establishing your own blog is easy and simple. You will surely encounter problems during the first few months of your site. However, you don't have to worry since there are a lot of resources you can use to solve your problems. There are also some online forums that you can join to get answers for your questions.

The Costs

These days, you can start a blog without spending any money. If you want to try how blogging works, it would be best if you will take advantage of the free blog hosting services currently available. This way, you will know whether you really want to be a blogger or not without wasting your funds.

Once you're sure that you want to be a blogger, you should host your own blog. People who want to earn money from their blog/s find more success by getting a hosting package for their site. Self-hosted blogs involve costs, but they are not expensive. In fact, you can get a self-hosted blog for just $15/month.

Chapter 2:

How to Earn Money Through Your Blog

T his chapter will teach you how to monetize your blog. It will provide you with tips, tricks, and techniques that you can use to start earning money from your new site.

The Techniques

CPC (i.e. Cost Per Click) Ads

A lot of bloggers monetize their sites using Cost Per Click ads. As its name suggests, a CPC ad system pays you each time a guest clicks on an ad. Adsense, one of Google's advertising programs, is the most popular CPC option nowadays. With Adsense, Google will check your articles. Then, it will search for advertisements that are related to those articles. Since the ads are related to the actual blog entries, this advertising technique produces excellent results.

Important Note: You will learn more about Adsense later in this chapter.

This system offers a lot of benefits to bloggers and readers. It helps bloggers to earn money from their online articles. Meanwhile, it helps readers find products and/or services that they need.

CPM (i.e. Cost Per Thousand) Ads

This kind of ad system pays you based on the number of visitors who see the ads. The "M" in CPM stands for the Roman numeral that represents 1,000. You won't earn much from this ad system during the first few months of your blogging career. However, once your blog gets a large number of traffic, CPM can help you earn large amounts of money. Here are the most popular CPM networks today:

- Adbrite.com
- Pulsepoint.com
- Casalemedia.com
- AdClickMedia.com
- Technoratimedia.com
- Adify.com

Each of these networks has distinct pros and cons. It would be best if you will analyze each network before getting them on your blog. This way, you can make sure that the CPM network you're using matches your needs.

Affiliate Products

As a blogger, you can act as the intermediary between sellers and potential buyers. You may form partnerships with people or businesses who offer products and/or services that are related to your blog. Then, you will recommend those products/services through your blog posts. In this setup, you will earn money whenever one of your readers pays for the said products/services.

This advertising system turns you into a salesperson. However, unlike traditional salespeople, you can promote products and/or services to countless people and form partnerships with numerous sellers.

Here are some of the best affiliate programs that you can use for your site:

- Flexoffers.com
- LinkShare.com
- Shareasale.com
- CJ.com
- E-Junkie.com
- Affiliate-program.amazon.com
- Panthera.com
- LogicalMedia.com
- RedPlum.com
- MoneySavingMom.com
- Coupons.com
- MySavings.com

It would be best if you will be honest with each of your recommendations. Give objective opinion about the products and/or services you display on your blog. For instance, you may create an article that lists the pros and cons of a product to educate your readers regarding that market offering.

This system works well because it helps three parties simultaneously. You earn money for your referrals. Sellers get more customers. Your readers, on the other hand, learn about the products/services that they might need. You will learn more about affiliate advertising later.

Ad Space

You also have the option to offer ad space to online marketers. Countless bloggers have tried this system and succeeded. Selling ad space to marketers is highly effective when used in niche markets. To apply this ad system on your blog, just visit the www.buysellads.com website. That website helps marketers and bloggers get connected.

The Timeframe

Now that you know how to monetize your blog, you probably want to know how long it would take for you to earn a consistent stream of income. It's important to point out that blogging is not the fastest way to earn money. It involves a long timeframe. Even the best bloggers needed to wait months (or even years) just to get a satisfactory level of income.

Obviously, you can shorten the timeframe significantly if you are already familiar with advertising, content creation, search engine optimization, and other topics related to blogging.

How to Use Adsense Effectively

This part of the book will provide you with advanced techniques. These techniques are designed to boost your earnings from Google's

Adsense program. By applying these techniques on your site, you can double (or even triple) your income from Adsense.

It's important to point out that each blog is unique. Even blogs that belong to the same niche can have different layouts, readers, and articles. These elements greatly influence potential earnings from the Adsense program. However, the techniques given below can help you with your Adsense campaign regardless of your niche, layout, and current blog entries.

1. Place ads on places that attract the readers' eyes – Keep in mind that you earn money from Adsense each time a reader clicks on the displayed ads. That means you need to place those ads in the attractive parts of your blog. However, you also need to consider the overall usability of your site. If you will place Adsense ads with reckless abandon, readers might stop visiting your blog. To gain maximum benefits from Adsense, you need to master the art of ad placement.

2. Target specific parts of your blog entries – With this technique, you will pinpoint the exact parts of your articles that Google must check when choosing ads. Implementing this technique on your blog is simple and easy. You just have to type "<!-- google_ad_section_start →" to indicate the place where Google should start checking. Then, use "<!-- google_ad_section_end → to indicate the endpoint for Google's analysis.

3. Use the plugin called "Quick Adsense" - This plugin, which is offered by the WordPress system, lets you incorporate

Adsense ads into your articles. With this tool, you can choose different criteria to customize the placements of your ads.

Important Note: According to experienced bloggers, the best spot to place an ad is right under the title of a blog entry. You can use Quick Adsense to place ads on that spot.

4. Blend colors – When the Adsense program was introduced, bloggers used loud colors to make ads more interesting. These days, however, bloggers change the colors of Adsense ads so that they blend with the site's theme. The main disadvantage of the old strategy is that it destroys the overall beauty of the blog. The newer strategy resolves this problem by making sure that the blog and the ads are pleasing to the readers' eyes.

5. Create articles for online searchers – You probably have regular readers and visitors from search engine results. In most cases, bloggers don't exert much effort on meeting the needs of visitors who used search engines.

It is true that most of these visitors won't return to your blog again. You need to embrace this fact if you want to succeed as a blogger. Keep in mind that you created a blog to share helpful articles with other people. Thus, you shouldn't worry about the number of times your readers visit your blog.

Search for the keywords used by online searchers. Then, incorporate those keywords to your articles. It doesn't mean that you will ignore your regular readers. Rather, you will write materials that can satisfy the needs of anyone who will visit your blog.

6. Install the search bar of Google – Google is one of the leading search engines today. Its search capabilities are stronger than that of any blogging platform. By installing Google's search bar on your blog, you can help readers find the information they need and earn some money. The results of the online searches come with standard ads, just like typical search results of the Google engine.

7. Connect Adsense and Google Analytics – By linking Adsense and Google Analytics, you can obtain loads of data about your earnings. This technique will help you determine your most profitable articles and best keywords. It can also pinpoint the third-party websites that send the most volume of traffic to your blog. As an Adsense user, you should take advantage of this option.

How to Earn More Money Through Affiliate Advertising

Bloggers consider affiliate advertising as one of the simplest sources of income currently available. It requires you to promote other people's products. With affiliate advertising, you will only earn money if a reader performs a certain action (e.g. buy something, register for an account, download an eBook, etc.).

To earn money through this channel, you must do the following:

1. register for one or more affiliate programs
2. encourage readers to do the action required by the advertiser
3. get paid each time a reader performs the required action

For example, let's say you promoted an eBook that costs $2. Because of the affiliate program, you will get $1 each time one of your readers downloads that eBook. Thus, if 50 people purchase that eBook, you will get $50 as "affiliate income."

In most cases, affiliate programs involve products that must be sold. However, there are also some programs that don't require purchases. For instance, if you are an affiliate of www.coupons.com, you will just post a coupon from that website. You will get paid each time a reader prints that coupon. This kind of program pays up to $0.80 per printed coupon.

Now that you know how affiliate advertising works, it's time to discuss how you can earn loads of money from it.

Here are some tips that can help you maximize your affiliate earnings:

- Consider your values – Affiliate advertising can serve as an excellent source of income. However, it involves potential problems that you should know about. For example, if the product you're promoting has poor quality, your reputation might be ruined. That means you shouldn't promote or write positive reviews about a product just because you can earn money from it. Before crafting any post regarding an affiliate program, ask yourself whether you will write that post even if you won't get anything in return. This way, you can make sure that you will give objective opinions regarding the product you're promoting.

- Use your posts for affiliate ads – You need to incorporate affiliate ads into your blog entries, instead of placing "affiliate links" on your blog's sidebar. Prior to writing a post about a product, find out whether the seller offers an affiliate program. This is an excellent way for you to increase your earnings.

- Try new stuff – Learning how to implement affiliate marketing on your blog requires time and patience. It's possible that you will experience failures during the first few months of your blog. However, don't be discouraged. Try new techniques, change how you promote affiliate links, and identify the methods that work for you.

- Be honest – Give honest opinions about the affiliate products you are promoting. It is also important to inform your readers whenever you promote other people's products. This way, your readers will know that you are getting something each time the promoted product is purchased. Online readers appreciate honesty – countless bloggers have received positive feedback from their readers because of this simple strategy.

- Don't place excessive affiliate links on your site – Don't sign up for each affiliate program you will encounter. It would be best if you will choose affiliate programs that match your blog's "vision" and "mission." Consistency plays a crucial part in the world of blogging. Make your readers feel that your posts and the products you are promoting are related.

Chapter 3:

How to Attract Visitors

Most people create a blog so they can share their thoughts and/or experiences with others. Unfortunately, readers won't visit a blog unless they know that it exists and that it contains useful information. To become a successful blogger, you need to know how to generate traffic for your site.

This chapter will teach you how to attract readers. You need to read this material carefully since it can help you maximize your earnings from your blog.

Comment on Other Blogs

This is one of the best techniques that you can use. Here, you will just place thoughtful responses on other people's blog posts. Great comments can drive a lot of visitors to your site. Bloggers are usually social – being active in the "bloggers' community" can be extremely rewarding. Aside from giving you extra visits, this technique will help you establish relationships with other members of the community.

Post on Other Sites

With this technique, you can get hundreds (or thousands) of readers quickly. Bloggers have used it to boost the traffic of their sites.

This technique requires you to write great articles and submit them to other websites. It's important to note that you should submit unpublished articles only. Other websites won't benefit from materials that have already been published. However, there are websites that allow you to republish your articles (you will learn about this later).

Submitting your work to other sites is easy. You just have to look for blogs that belong to your niche. Comment on the posts inside those blogs and befriend their owners. Finally, tell them that you would like to submit unpublished articles.

Join Blog Networks

You can boost the traffic of your blog by expanding your blogging network. Two of the leading online communities for bloggers are www.blogcatalog.com and www.technorati.com. Join these communities and search for blogs that are related to yours. Building relationships with other bloggers will improve the overall traffic of your site. Additionally, establishing a wide network is one of the best things that you can do as a new blogger.

Join the Blogcarnival.com Website

Expert bloggers recommend this technique to newbies. Basically, www.blogcarnival.com is a website that allows you to share articles you've published before. That means you can use this site to promote the contents of your own blog. Blogcarnival.com can help you get a lot of traffic, especially if your articles have interesting titles.

Join Blog Directories

Inexperienced bloggers need to look for blog directories that are related to their topic. This technique is so important that you must do it before writing blog entries. Here, you will just access your favorite search engine, type your chosen topic, add the word "directory," and hit the Enter key. Your screen will show you blog directories that are relevant to your chosen subject.

As a general rule, you should stay away from directories that offer 100% acceptance. Search for directories that will check your blog before approving it. This technique will help you ensure that you are dealing with topnotch directories. Here are some of the best directories that you can join:

- www.joeant.com
- www.blogged.com
- www.botw.org
- www.greenstalk.com
- www.dmoz.org
- dir.yahoo.com
- www.familyfriendlysites.com

Comment on Message Boards

You can increase the number of visitors to your site by being active in message boards or online forums. Place the URL of your site on your forum account's signature line. Run online searches for forums that are related to your blog.

Here's a basic rule that you should remember: Make sure that all of your posts on the forums are genuine and on-topic. If you will try to promote your site in each of your posts, other bloggers will just get annoyed. They will know what you're trying to do, and your plan will surely backfire.

Create Pages on Other Sites

There are some sites (e.g. www.hubpages.com, www.squidoo.com, www.infobarrel.com, etc.) that let you create a page about any topic. Establishing this kind of site is easy and simple. As an added bonus, this kind of page usually gets excellent search engine rankings.

If this "basic page" acquires lots of traffic, your main blog will also gain more visitors. You can also monetize your "basic page/s", which means you will have more opportunities to earn through your online articles.

Chapter 4:

SEO and Blogging

You might need to wait a long time just to get sufficient traffic from search engines (e.g. Google, Yahoo!, Bing, etc.). In general, search engines are "unfriendly" when it comes to new sites. Search engines prefer established and reputable websites. This is the reason why you need to get associated with legit directories, bloggers, and sites.

Your site will benefit from search engines once it has a lot of links and pages. That means you just have to be patient while expanding your site. Continue writing excellent content – after some time, you will be able to reap the benefits offered by search engines.

In this chapter, you will learn about the different SEO techniques that are compatible with blogs. Read this chapter carefully since it will aid you in boosting your blog's overall traffic.

SEO – The Basics

SEO (Search Engine Optimization) helps websites in getting excellent rankings in the search engine results. Your website will get a lot of free traffic if it can obtain great rankings for related keywords. This book won't give detailed explanations of all the SEO techniques currently available.

Rather, it will focus on the basic techniques that inexperienced bloggers can use. Keep in mind that you don't need to be an SEO expert to earn money from your blog. You just have to understand how SEO works and how you can use it for your site.

Search engines are designed to help searchers find the content, products, and/or services that they need. These "engines" want to show websites that are relevant to the searchers' needs. The system used by these engines aren't perfect. However, Google and other search engines are continuously working to improve the accuracy of their results.

Important Note: Never try to trick search engines. Your site will get punished once these engines catch you. Rather than finding and exploiting the technical issues of search engines, you should focus on writing content that can educate and/or entertain your readers.

The SEO Techniques

This part of the book will arm you with fundamental SEO techniques. If you will use these techniques properly, your site will get excellent rankings in the search results pages.

- Use Proper Title Tags – Bloggers consider this as one of the most important SEO techniques.
  Basically, title tags are the words/phrases/sentences that appear at the top section of a browser. Search engines consider the title tags of a website while determining its rank.

- Include keywords in the title tags of your webpages. In addition, you need to make sure that your title tags are related to your content. For instance, if you wrote an article about chess, you don't want to use "Best Basketball Drills" as the title tag for that particular entry.

- Incorporate Keywords to Your Anchor Texts – An "anchor text" is the word or phrase you will use when creating a link. For example, if you will place your mouse pointer over this link, you will see that it points to the Yahoo! website. In this example, "this link" serves as the anchor text. This element of content creation is important because it gives a lot of information about your site.

- Search engines consider anchor texts when ranking websites. As you know, it's not possible to manipulate how other bloggers create anchor texts when linking to your website. However, you should incorporate relevant keywords to your anchor texts whenever you can. For example, you may use keyword-rich anchor texts when linking to the posts and pages within your blog. You may also use this technique on your directory submissions and forum signature lines.

- Choose Keywords Carefully – If you want to earn money through your blog, you should identify what potential readers are searching for. You can maximize the benefits of this technique if you will do it before writing any article. There are numerous keyword tools available online. Use these tools to find the hottest keywords that are related to your chosen topic.

- Get Inbound Links – Inbound links play an important role in any SEO campaign. Acquiring links that point to your site is one of the most effective techniques for improving a website's search engine rankings. As a bonus, inbound links can send potential readers straight to your blog. Many bloggers overlook this fact. Your blog will enjoy a spike in overall traffic if popular blogs or sites point to it.

The Tools

In this part of the book, you will discover the tools that can boost your site's traffic and search engine rankings.

- www.bigstock.com - This is one of the best websites that offer stock photography. With www.bigstock.com, you can get great images without spending much money. In general, articles become more attractive and pleasing to the eye when they contain images. Thus, if you want your readers to enjoy your written work, you should add pictures to them.
- www.google.com/webmasters/tools – This tool shows you how Google sees your blog. It will also pinpoint the problems that are affecting your blog's rankings in the search results pages.
- www.google.com/analytics – This powerful tool helps you in tracking and analyzing your readers. Google has integrated this tool with Adsense. Thus, it can now provide you with more useful information.
- www.mailchimp.com – Email marketing can help you attract more visitors. If you just want to try email marketing, www.mailchimp.com is the best tool for you. With this tool,

you can get up to 2,000 email subscribers without shelling out any money.

- tools.seobook.com/keyword-tools – You can use this tool to determine how many online searchers use certain keywords. Keywords play a crucial part in your blog's SEO campaign so you should add this tool to your arsenal.

Chapter 5:

Writing Content

You won't earn money from your blog if you can't write great content. That's because you need sufficient traffic in order to earn income from your website. As you probably know, nothing beats great content when it comes to attracting website traffic.

If you will ask experienced bloggers for advice on how to succeed, almost all of them will tell you to write excellent content for your site.

In this chapter, you will learn the basics of high-quality content. It will also provide you with tips and techniques that can help you write effective blog posts.

Write Articles that are Unique, Engaging, and Valuable

To become successful, you need to write unique content. If your blog has unique articles, readers will keep on visiting it. If you will just rewrite published materials, on the other hand, people won't visit your website. They'll prefer to read the original material.

Next, write articles that are engaging. The main reason why blogs gained immense popularity is the fact that they allow readers to interact with the writers. Thus, as a blogger, you won't be having any monologue (unless no one reads and reacts to your materials).

People will visit your site, read the stuff you wrote, and react through comments or emails. That means you will interact with your readers actively. In order to create a well-known site, you must establish an online community for it. Here are the things you can do to establish a great community for your blog:

- Be personal, especially when writing your materials
- Ask your readers directly
- Write comments based on your readers' previous comments
- Share your ideas and opinions only during appropriate situations

Lastly, your content should be valuable. If the materials you will publish don't have any value, no one will visit your blog. The value of your materials may differ, based on the topic you're writing about. For instance, if your blog focuses on tech news, you need to craft blog entries that inform your readers about the latest technological developments. If your blog focuses on humor, however, you must create posts that make your readers laugh.

Write Killer Articles

Basically, killer articles are blog entries that contain exceptional value for the readers. Whenever a reader encounters a killer article, he/she will blog and/or Tweet about it, bookmark it, and inform other people about the article's existence.

Often, a killer article is structured and contains a lot of words. However, you're not required to use thousands of words in order to create killer articles. Let's assume that you have a chess blog and that you discovered a new strategy to destroy White in the Ruy Lopez opening. This kind of topic will surely produce a "killer," even if you will keep the article concise. That's because the article offers enormous value for your readers (considering that the Ruy Lopez is one of the most popular openings today).

As a blogger, you should consider killer articles as the foundation of your content development scheme. This way, you can generate website traffic, attract inbound links, and present yourself as an authority in your chosen niche. Killer articles may take any of these formats:

- Rankings (e.g. Top 10 Chess Players of the 21st Century)
- Lists (e.g. The 5 Opening Blunders You Must Avoid)
- Breaking News (e.g. Magnus Carlsen Defends His Crown for the 20th Time)
- Resources (e.g. Free Chess eBooks You Can Download Today)
- Elaborated Interviews (e.g. Ten Chess Opening Experts: The Downfall of the Semi-Slav Opening)
- In-depth Analysis (e.g. Your Detailed Guide to the Caro-Kann Opening)

Killer articles require you to perform researches and analyses. In addition, these kinds of article are more complex than ordinary ones. Fortunately, the results you can get from killer articles are definitely worth it.

Lastly, publish killer articles on a regular basis. It would be great if you can publish a killer article each week. If your schedule doesn't permit that, however, try to create a killer article once per month.

Regular Posts

It is true that killer articles will attract new readers and establish your authority in the niche you are in. However, you can't rely on killer articles all the time. You also need to write regular blog entries to create an online community for your site. Regular posts let you interact directly with your readers.

In addition, ordinary posts can help you ensure the smooth flow of your blog's contents. Creating killer articles on a daily basis would require lots of effort. Achieving this unlikely goal wouldn't benefit your blog in anyway – it will just overload your audience with complex and disconnected bits of information. Here are the most popular variants of ordinary posts:

- Polls (e.g. Who is Your Favorite Chess Player?)
- Events (e.g. Chess World Championship 2016)
- Site Updates (e.g. The Latest Features of www.thechessblog.com)
- Quick Links (e.g. 10 Tips for Sicilian Defense Fanatics)
- Opinion Pieces (e.g. Garry Kasparov's Greatest Strengths)

Headlines

You need to know the importance of headlines if you want to be a successful blogger. Keep in mind that you need to find the best headline for each of your blog posts. Actually, some experienced bloggers suggest that the amount of time you will spend on finding a headline should be equal to the time you spent on writing the article itself.

The headline is important because it attracts your readers. There are countless blog entries on the internet today – readers choose the ones they want to read based on the articles' headlines. Readers look at the headline of your article before reading its content. If your headline sucks, readers might ignore your article completely.

Headlines affect your current and potential readers. For example, if your post has a bad headline, even your regular readers will skip it. The time available to online readers is severely limited. They won't waste their time on materials that are "probably useless." Additionally, if you will submit that article to social bookmarking websites (e.g. www.digg.com), readers won't click on it.

Effective headlines have two main elements, which are:

1. Wording – The headline's wording should match your readers' mindset. This element informs potential readers that the article they're looking at contains the information they need. Here, knowledge regarding keyword selection can help you greatly.

You may use Google's "Keyword Tool" for this. Basically, "Keyword Tool" is a service that generates 150 words related to the keyword you will enter. This tool will also display the search volume of each related term. For example, if you will launch the Keyword Tool and type "energy," your screen will show you this:

Write Down Your Ideas

Bloggers usually have problems in generating new ideas. You can solve this problem by writing down your thoughts. Write down all of your ideas that may lead to blog entries. You may use any tool for this process (e.g. a notebook, a text editor, a whiteboard, etc.). It doesn't matter what tool you use. If your tool lets you store, organize, and retrieve your ideas, you're good to go.

As an alternative, you may create a draft whenever you get an interesting idea. Just launch your preferred blogging program and record your thoughts. Type down the entry's title and main points. You don't have to finish the article in one sitting. Save the entry as a draft (if you can't complete it now) and work on it again once you have the time for it.

If the strategies given above are not enough to generate a consistent stream of ideas, you may launch Google's Keyword Tool and enter some keywords that are relevant to your topic. Find the long versions of the keywords and use them as the basis for your future articles.

Lastly, you may visit social bookmarking websites. This strategy will help you discover the hottest topics on the internet. Here are the leading bookmarking sites today:

- www.delicious.com
- www.digg.com
- www.reddit.com
- www.stumbleupon.com

The Frequency of Your Posts

There is no such as thing as "ideal frequency" for publishing blog entries. Some blogs get updated once per week.

Other blogs, however, display several new posts on a daily basis.

It is important to point out that quality is more important than quantity. Thus, you need to make sure that all of your posts are useful for your readers and your website. Refraining from posting new articles is better than posting bad materials. The absence of new posts requires your readers to wait. The presence of low-quality posts, on the other hand, can ruin the image and popularity of your blog.

To earn money from your blog, you should strike a balance with quantity and quality. This is the reason why most of the greatest blogs today publish new entries daily. If your schedule doesn't let you post every day, try to be consistent regarding the time you publish blog entries. For instance, you can publish three new posts each week and follow that schedule.

To maximize the benefits you can get from being consistent, try to publish new content on the same day/s each week. This way, you can assure your readers that they will find new entries on your site.

Five Tips for Crafting Excellent Posts

1. Be true to yourself – Inexperienced bloggers tend to imitate the writing style of other people. They do this in order to display a false air of confidence and/or authority. Unfortunately, this tendency can lead to complex problems in the future. Rather than trying to imitate other blogger's style, you should try to find your own "writing voice" and improve it.

You have unique characteristics. Your skills and abilities are different from those of other bloggers. You view things differently, because you are different from others.

Study the techniques used by other bloggers. This approach allows you to enhance your writing skills quickly and easily. However, keep in mind that you are different from them.

2. Show that you are confident – New bloggers often get discouraged whenever they see other blogs that have a lot of traffic, updates, creative posts, social media followers, or comments. These

inexperienced bloggers worry too much about failing to emulate the achievements of their "seniors."

You won't benefit from worrying or sulking in a corner. If you're not happy with your current skills, do what you can to improve them. Try to show your confidence in your blog posts. Posts that were confidently written have high chances of turning visitors to regular readers.

3. Engage your readers – Communicate with your audience. Reply to their emails and comments when you have the time. Take note of your readers' input and advice. Make your readers feel that you value them. This way, you can strengthen the loyalty of your readers.

Spice things up by publishing off-topic photos, videos or posts. These "surprising" blog entries can keep your blog exciting. Additionally, you must try new blogging techniques. Writing posts using the same technique over and over again can result to bored readers and poor website traffic.

4. Be honest – You don't need to be perfect in order to write great stuff. In blogging, honesty is more important than apparent perfection. Don't give the impression that you know everything and that all of your actions are correct.

You can attract a lot of readers just by being open and honest. In some cases, you also need to show your weaknesses.

5. Review and edit your output thoroughly – Readers don't like typos and grammar mistakes. If you want to earn money through your blog, you need to enhance your writing and proofreading skills.

Sloppy writing can ruin your plans. You can secure the quality of your blog by proofreading each post twice prior to publishing it. Work on your writing skills constantly. Read excellent blogs and books. Be objective while reviewing your work.

Here are some tips that can help you write great content:

- Don't write huge paragraphs – Online readers get turned off by huge paragraphs. As you probably know, reading large blocks of characters on a monitor is not fun. You can enhance the readability of your posts by reducing the size of your paragraphs.

- Be consistent regarding the font you're using.

- Don't use ellipses, all caps, or exclamation points unnecessarily.

- If your post contains multiple paragraphs, use subheadings.

- Add photos and/or videos to your posts.

Chapter 6:

Site Design and Functionality

T his chapter will teach you how to improve your site's design and functionality aspects. Study this material carefully: your site's look and usability greatly affect the income you can earn from your written works.

The Free Stuff

Inexperienced bloggers must not spend a large amount of money on their blog's design. Most blogging platforms offer beautiful themes for free. For example, WordPress has various themes you can choose from. You just have to log in to your WordPress account and access the page called Themes Directory.

Additionally, there are blogs and websites that offer lists of excellent themes. Launch your favorite web browser, type the right keywords (e.g. "best themes wordpress"), and hit the Enter key. In just a few seconds, your screen will show you list of the best themes currently available.

You can easily customize the free themes offered by WordPress. Thus, you can personalize your blog completely even without spending any money.

You can also benefit from studying basic CSS and HTML. These topics can help you improve the look of your blog. Currently, there are countless eBooks and online articles written about CSS and HTML. You will get all the resources you need just by running an online search.

Your Blog's Logo

If you have some money for your blog's design, use it to get a logo. That's because you can use the logo of your blog on any theme, design, or template. It will also strengthen your brand and improve the uniqueness of your blog.

Currently, www.99designs.com is the leading website when it comes to purchasing logos. With this site, you can

obtain an excellent logo   for just $150.

Implement Prioritization

Your readers can do a wide range of things while on your blog. Here are some examples:

- read one post
- read a lot of posts
- read the "most viewed" posts
- subscribe to your site's RSS feed
- click on an ad

- purchase products through the affiliate links on your blog
- sign up for your email newsletter
- add a comment
- download an eBook
- buy a product you're selling
- share one or more posts with others

However, there is no blog design that can encourage all of the activities listed above. Many bloggers have tried, but without success. These bloggers succeeded in filling their sites with clutter, confusing their readers, and reducing their blog's effectiveness.

That means you must prioritize certain user activities over others. It would be best if you will list down the five user activities that are important for your blog. Then, design your blog so that it emphasizes the activities you listed down. If a design element doesn't match your priorities, you should either remove it or transfer it to a different part of the website.

Encourage Subscriptions

The previous section required you to list down the most important activities for your site. Well, encouraging email and RSS subscribers should be at the top of your list. That's because registered subscribers often become regular readers. In addition, you can easily establish relationships with readers who view your content on a regular basis.

Most bloggers support user subscriptions through RSS feeds. This option is good, but you shouldn't rely on it exclusively. To improve your chances of succeeding, offer email subscriptions too. If you have a Twitter account, you may also ask your readers to follow you on that social networking site so they can get updates easily.

Your "Bestseller"

While designing your blog, you should also strive to display your best posts. This way, new visitors can easily find the best articles your blog has to offer. By showing excellent contents to your first time readers, you will have better chances of converting them into regular visitors. They might also sign up for your email and/or RSS feed subscription offers.

You can display your best materials in two ways. First, you can create a "best material" area within your blog. Most bloggers choose the header, footer, or the sidebar for this purpose.
To list your best articles, you may use HTML codes to add the links manually. If you want to automate this task, on the other hand, you may use a plugin offered by your blogging platform.

For the second option, you will create a webpage inside your blog to showcase your best content. Here, you may divide your posts based on their category or date of original publication. After creating the webpage, you may display its link on your blog's sidebar or main menu. You may boost the appearance of your blog by replacing the link with a relevant image.

The Mistakes Related to Usability

This part of the book will discuss the most common usability mistakes committed by inexperienced bloggers. Read this material carefully – it will help you provide excellent user experience to your site visitors.

- Old records are hard to find – Your blog should have a page that contains all of your previous posts. Adding an archives section to your blog helps your readers in finding the material they need. It can also enhance your rankings in the search engine results pages.
- Search boxes aren't available – People use search boxes in order to find the information they need. Some individuals even use a search box to reach different parts of a website. Your users might get frustrated if your blog doesn't have any search box.
- The site's structure is too complex – As a blogger, you need to keep your site's structure simple. Here are some things that you can do to improve the structure of your blog: (1) eliminating/minimizing drop-down options, (2) adding a navigation bar, and (3) making sure that all of your blog's webpages have a link that points to its homepage.
- Poor typography – Your readers will stop visiting your blog if they have problems reading your content. Use the proper font size for your blog entries. Ensure that the spacing between lines and letters are enough. Lastly, make sure that your webpages contain sufficient amounts of "white space."

- Excessive advertisements – Some bloggers think that they need to place numerous ads on their site just to make money. Well, nothing could be further from the truth. You will only make money from blogging if you have great website traffic. Unfortunately, readers don't like excessive ads. That means placing lots of ads on your blog can actually diminish your site's traffic and your overall income.

  It would be best if you will start with a few advertisements. Increase the number of ads slowly. Make sure that this "monetizing" process doesn't affect the user experience received by your readers.

- Problematic links – Links play an important role in your site's navigation. In general, your readers should be able to recognize links easily. That means you should underline the hyperlinks inside your posts. If you're not fond of extra lines, however, you may change the color of the hyperlinks. This way, your readers can easily identify the clickable and non-clickable parts of your blog entries.

- Too much widgets, badges and buttons – Inexperienced bloggers usually commit this mistake. Adding a badge, a widget, or a button is easy and simple. As a result, bloggers use these tools to decorate their sidebars.

  The habit of installing unnecessary items on a webpage can hurt your site's traffic. Excessive buttons, widgets, and/or badges don't contribute in increasing your content's value. Rather, they can actually confuse and/or frustrate your readers.

Chapter 7:

The Networking Aspects of Blogging

These days, the people/organizations you are linked to have the same value as the things you actually do. That statement also applies to blogging.

If you can build great relationships with other bloggers, you will obtain a lot of backlinks. Additionally, these people will recommend your site to their readers and assist you in selling products and/or services. Simply put, networking can help you earn money.

This chapter will teach you how to conduct proper networking. It will provide you with tips and strategies to help you build an effective network.

Real Connections

When it comes to networking, you should always focus on establishing "real" connections with other people. Don't approach a blogger just because he/she is rich, important, or popular. Rather, establish relationships with the bloggers who have earned your respect. Find people whose materials you can gladly recommend to your own readers.

If you will follow the tip given above, you will be able to form win-win relationships with the people who can help you best. Keep in mind that networking involves the principle of mutualism: help others if you want them to help you. Be ready to support the bloggers who belong to your network.

List Blogs Down

Before establishing your online network, list down the blogs that belong to your chosen niche.

you may also include blogs that belong to niches that are related to yours.

If you want,

The number of blogs you have to list down depends on the niche you are in. If you are working on a tech blog, your list should contain about 200 items. If you're blogging about cockatoos, however, 10 to 15 blogs must be enough. While doing this task, you need to get all of the information available to you. This list will help you build your network. If your list is incomplete, your chances of succeeding will be extremely low.

To find blogs that belong to particular niches, use the following tools:

- www.alltop.com
- www.blogcatalog.com
- www.technorati.com

- www.blogrank.com
- www.wikio.com/blogs/top

Approach the Bloggers

After creating your list, you need to approach the owners of the blogs you found. Almost all blogs have an email address or contact form. Visit the blogs one by one and check their "Contact Us" page (if applicable).

The following tips will help you in introducing yourself:

- Be straight to the point. Say what you want to say.
- Tell them that you are blogging about the same niche/topic/industry. Include your blog's URL in your signature line or in the content of your message.
- If the content of their blog interests you, inform them. Additionally, tell them that you would like to link to their published materials if it's okay.

Link to Other Sites When Appropriate

You should link to other sites in order to get links for your own blog. Obviously, other bloggers will likely create links that go to your site if you will do the same for them first. As you can see, the "golden rule" also applies to online networking.

It would be great if you will subscribe to the RSS feeds of the blogs you listed. With this trick, you will keep yourself updated regarding the hottest trends in your chosen topic. In addition, it will help you find great articles from the blogs you have subscribed to. Whenever you find an awesome article, write your own blog entry based on it, say whatever you want to say regarding the topic, and indicate the URL of the original material.

You may also contact that blogger through an email. Tell him/her that you like his/her work and that you shared it with others.

Lastly, make sure that you are linking to high-quality posts that are related to your niche. Great articles won't help you if there's no connection between them and your blog. Thus, if you're blogging about chess, linking to a post about the newest BMW cars isn't a good idea.

Support the Members of Your Network

Aside from creating outbound links, you can support the people inside your network in many ways. Here are some examples:

- recommend their sites to your friends and readers
- promote their products and/or services
- recommend them for awards or interviews
- share and vote for their blog entries on bookmarking websites
- share their content on social media platforms such as Facebook and Twitter

Important Point: Other bloggers will consider you as a friend if you will continue "giving" without asking for anything in return.

Promote Your Best Posts

The tips and strategies given above focus on doing things for other people. You're probably thinking how you can reap the benefits of having an online network. Well, if you will follow the lessons previously discussed, other bloggers will notice you and promote your blog in return.

This process might take a long time. Fortunately, there's a technique that you can use to encourage other bloggers to link back to your site. This technique requires you to share your blog's "bestsellers" to the bloggers you are trying to attract.

For example, after publishing a killer article, you may send the URL of that article to your target bloggers. While using this technique, you should never beg for backlinks. Just inform the other bloggers that you wrote an article that they might like. Then, provide a link that points to that article.

This approach doesn't pressure the recipients. Thus, they have higher chances of reading, reviewing, and recommending the post you shared. Keep in mind that other bloggers might not give you the backlink you're waiting for. In this situation, stay calm and concentrate on what you need to do. If you will keep on improving your blog and your writing skills, those people will surely link to your posts.

Chapter 8:

How to Earn More Money

This chapter will help you increase your earnings. It will give you tips and techniques that you can use to earn more from your blog.

Monetizing Your Sidebar

Offering sidebar advertisements require a lot of time and effort. However, as numerous bloggers have discovered, sidebar ads can boost blog earnings significantly. In addition, you can implement this scheme regardless of the niche you belong to. By offering sidebar advertisements on your own, you will get full control over the ads that will appear on your blog.

Here are the things that you can do to monetize your sidebar:

- Inform Others – Let other people know that you are offering sidebar ads. Many bloggers wonder why no one buys the sidebar advertisements they are offering. Well, the main reason is that these bloggers don't market their offerings actively. Online marketers won't pay for your sidebar ads if you won't inform them about your offering.

Marketers are usually busy – they won't analyze blogs and see whether the owners offer sidebar ads. Actually, some marketers won't know that you are selling sidebar advertisements unless you indicate it clearly.

Create an advertising page. Then, access your header and add a tab that points to the advertising page you created. Basically, your advertising page must show the following details:

- the demographics of your readers

- testimonials from current and/or previous marketers
- the daily and/or monthly traffic of your blog
- advertising options and prices

Your blog's advertising page must also state why advertisers should choose your site. Let potential advertisers know what they are going to get from your sidebar ads.

- Offer Discounts – If it's your first time selling sidebar advertisements, you may offer discounts to attract potential customers. Create a blog entry that highlights the reduced pricing. Then, forward that post to the companies who might be interested in your sidebar ads.

- Create "Ad Bundles" - Offering basic sidebar ads can generate income. However, you can boost your earnings further by creating "ad bundles." For example, rather than offering a sidebar advertisement for $25/month, offer a two-month "bundle" that contains one sidebar ad, one post about the customer's business, and one link on your email and/or RSS feed for just $70.

Alternatively, you may offer price reductions to people who buy ads that last for several months. This kind of discount leads to a win-win situation for you and your customers. They will be able to reduce their marketing costs, while you won't have to worry about finding new advertisers each month.

- Fill Your Ad Spots – According to experienced bloggers, you shouldn't leave your sidebar ads empty. Don't place an "Advertise Here" sign on your blog. This kind of notice tells potential advertisers that no one wants to pay for your sidebar ads. Obviously, you don't want your potential customers to think that way.

If there's a free space in your ad spots, use it to promote affiliate products. Alternatively, you may offer free advertising to your friends.

The Pricing

The price that you will charge for your sidebar ads depends on various factors. These factors are:

- the traffic of your blog
- the niche you belong to

- the placement of the ad
- the number of ad spots you're offering
- the current demand for ad spots

Beginners should start with low prices. Increase the amount as you gain more blogging experience and site traffic.

Usually, you will get paid per thousand views. That means you should start with $1 or below per thousand views. If you are currently getting 15,000 views each month, you may get about $15/month for a 250x250 ad that is placed on your sidebar.

Important Note: Don't sell more than eight sidebar advertisements. The overall value of your ads will decrease if you will place a lot of sidebar ads on your blog. In general, few high-paying ads are better than numerous yet cheap ones.

Chapter 9:

Ad Networks

A d networks can serve as a great source of income for bloggers. Basically, an ad network is an advertising broker. You will reserve ad spots on your site and the brokers will sell the spots on your behalf. These brokers will get a fee (30-50% of the total price) each time they successfully sell an ad spot.

Many bloggers love ad networks. On the other hand, there are some bloggers who had horrible experiences because of this kind of network. In this chapter, you will learn the advantages and disadvantages of joining ad networks.

The Advantages

- Ad Networks Can Simplify Complex Tasks – Selling advertising space personally requires a lot of time and effort. By joining an ad network, you can simplify your tasks greatly. With this kind of network, you will just sign the contract, insert HTML codes into your sidebar, and collect the payments.

- Ad Networks Can Help You Earn More – The income that you will earn through an ad network varies widely. Some bloggers earn $1 to $2 per thousand views while others get $20 for the same number of views.

However, ad networks are still better than selling sidebar ads personally. The latter generates $1 (or below) per thousand views. Well, this amount is unacceptable for people who belong to an ad network. On average, network members get $2 to $4 per thousand views. That means joining an advertising network is better than offering ads personally, especially if you're an inexperienced blogger.

- Ad Networks Can Give You Extra Projects – Although the income from ad spots can be great, the income from side projects is often greater. Today, ad networks can provide you with extra projects such as sponsored posts.

Ad networks serve as the middlemen between bloggers and advertisers. Thus, they can give you excellent opportunities you won't find on your own.

The Disadvantages

- Ad Networks Control Your Deals – If you are a member of an ad network, you won't be able to choose which ads will appear on your blog. Most ad networks are not willing to make bloggers decide regarding the ads that will be displayed. This can be a problem – conflicts between your content and the ads you are showing might arise.

- Joining an Ad Network Is Not Easy – Joining an advertising network can take several months. Ad networks, particularly the best ones, have few openings and long waiting lists.

Chapter 10:
Additional Options

The previous chapters taught you how to earn money straight from your blogging activities. However, it is important to note that blogging allows you to earn income indirectly.

In this chapter, you will discover several strategies that you can use to earn extra income. These strategies use your blog indirectly.

- Conduct Online Classes – This strategy can help you learn new things while earning more money.

    Online teaching doesn't require a large capital. In fact, you just have to acquire basic tools (e.g. microphone). Countless bloggers have succeeded in implementing this strategy on their sites.

- Write eBooks – You can also earn money by creating and selling your own eBooks. This option is usually not included in a blogger's arsenal. Well, that's unfortunate. You can earn hundreds (or even thousands) of dollars annually just by writing and selling eBooks.

Writing eBooks can serve as your main source of income, especially during the first few months of your blog. Here are the things you need to keep in mind while using this strategy:

- Write about topics that are practical and relevant. Thus, you need to craft an eBook that teaches people how to do stuff. For example, you may write an eBook that explains how people can lose weight, start a business, become more organized, etc.

- Make sure that your eBook's cover looks amazing.

- Promote your eBooks through different marketing channels.

- Be a Freelancer – Some of your readers might offer you freelancing jobs. This strategy can be extremely profitable, especially if you have excellent writing skills. If you want to get more freelance writing opportunities, you may contact parenting magazines in your country. Most of them pay about $25 for each article.

- Be a Consultant – These days, some companies are hiring bloggers as consultants. Bloggers know a lot about online and social media marketing. Thus, companies are now trying to leverage bloggers' knowledge and experience.

Consulting services can get you about $50/hour. Unfortunately, finding clients can be difficult. Bloggers attract clients through their online networks.

Conclusion

Congratulations for finishing this book!

I hope it will help you earn income through your blog. By reading this book carefully and applying the lessons it contains, you will get regular readers and consistent streams of income.

The next step is to continue improving your writing and marketing skills. Write excellent content and share your materials with others. This way, you can become a blogger who helps his/her readers and gets paid for what he/she is doing.

Finally, if you enjoyed this book, please take the time to share your thoughts and post a positive review on Amazon. It'd be greatly appreciated!

Thank you and good luck!

CPSIA information can be obtained
at www.ICGtesting.com
Printed in the USA
BVHW080220020321
601389BV00002B/104